# *The U.S. Army Corps of Engineers: A History*

*Headquarters*
*U.S. Army Corps of Engineers*
*Office of History*
*Alexandria, Virginia*

2008

This is the Official U.S. Government edition of this publication and is herein identified to certify its authenticity. Use of ISBN 978-0-16-079585-5 is for U.S. Government Printing Office Official Editions only. The Superintendent of Documents of the U.S. Government Printing Office requests that any reprinted edition clearly be labeled as a copy of the authentic work with a new ISBN.

# Foreword

This illustrated history of the U.S. Army Corps of Engineers provides an overview of many of the missions that engineers have performed in support of the U.S. Army and the Nation since the early days of the American Revolution. A permanent institution since 1802, the U.S. Army Corps of Engineers has effectively and proudly responded to changing defense requirements and has played an integral part in the development of the Nation.

Engineers have served in combat in all of our Nation's wars. Throughout the nineteenth century, the Corps built coastal fortifications, surveyed roads and canals, eliminated navigational hazards, explored and mapped the Western frontier, and constructed buildings and monuments in the Nation's capital.

In the twentieth century, the Corps became the lead federal flood control agency. Assigned the military construction mission in 1941, the Corps constructed facilities at home and abroad to support the U.S. Army and the U.S. Air Force. During the Cold War, Army engineers managed construction programs for America's allies, including a massive effort in Saudi Arabia.

When the Cold War ended, the Corps was poised to support the Army and the Nation as we adapted to the new era. But the events of September 11, 2001, changed the diplomatic and military climate dramatically. After supporting recovery efforts at the World Trade Center and the Pentagon, Army engineers played an important role in the rapidly evolving Global War on Terrorism. Following combat operations in Afghanistan and Iraq, the Corps of Engineers established new overseas districts and a division in those countries to help rebuild their shattered infrastructures and foster a new era of peace and democracy in the region. The results of that effort will shape the Nation's future in the twenty-first century.

Today, building on its rich heritage, the Corps is changing to meet the challenges of the future. The U.S. Army Corps of Engineers is a broad-ranging engineer force of highly qualified civilians and Soldiers, working with our partners to deliver innovative and effective solutions to the Nation's engineering challenges. We are a values-based organization, focused on our mission and those we serve, dedicated to public service, and a vital part of the U.S. Army.

Our mission areas include planning, designing, building, and operating water resources and other civil works projects; designing and managing the construction of military facilities; providing immediate and long-term support to the public during natural disasters and national emergencies; and offering design and construction management capabilities for other Defense Department and Federal agencies and for foreign countries.

I hope that readers of this history will gain an appreciation of the military, political, economic, and technological factors that shaped the modern U.S. Army Corps of Engineers. We in the Corps, both Soldiers and civilians, are proud of our many past contributions and look forward with confidence to continued service as a relevant, ready, responsive, and reliable organization, proudly serving the armed forces and the Nation.

R. L. VAN ANTWERP
Lieutenant General, US Army
Chief of Engineers

# Acknowledgments

This publication is a revised and expanded version of *The History of the U.S. Army Corps of Engineers* (Alexandria, VA: Office of History, Headquarters, U.S. Army Corps of Engineers, 1998), EP 870-1-45. William C. Baldwin was project manager of this publication. Over the years numerous present and former Army engineer historians have contributed to the text. The historians who worked on this publication were Paul K. Walker, chief of the Office of History; William C. Baldwin; John Lonnquest; Matthew Pearcy; Michael Brodhead; Eric Reinert; James Garber; Martin Reuss; and Kent Sieg. Larry Roberts, Charles Camillo, Charles Parrish, Joan Kibler, and the USACE Command Sergeant Major, Robert Winzenried, also provided helpful information. Jean R. Diaz, the Office of History's editor, edited the text and managed the images. Janet R.M. Fitzgerald, BFA, Visual Information Specialist, Humphreys Engineer Center Support Activity, designed the cover. Douglas J. Wilson assisted in the proofing, and our administrative officer, Anna Punchak, assisted with financial and other matters. EEI Communications, Inc., under the team leadership of Jayne Sutton, designed and prepared the book for printing.

The illustrations for this book came from the Research Collections of the Office of History unless another source is noted below the caption.

# Contents

## Introduction



## 1. The Making of the Corps



## 2. Contributing to National Development



## 3. Strategic Role in War and Peace



## 4. A Vital Part of the Army



## Appendices

# Historical Time Line 1775–2005

## 1775–1815

**1775 ▶**
Congress established the Continental Army with provision for a Chief Engineer (June 16). Richard Gridley named first Chief Engineer and oversaw fortification at the Battle of Bunker Hill.

*Col. William Prescott at the Battle of Bunker Hill,* painting by Frederick C. Yohn
*The Continental Insurance Company*

**1779**
Engineer officers and companies of sappers and miners formed into a Corps of Engineers.

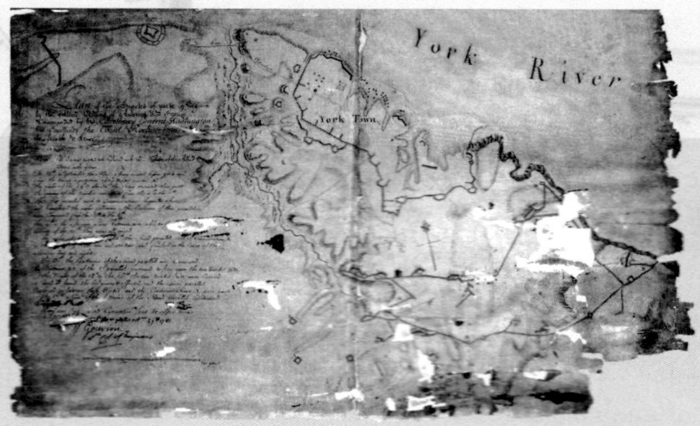


Plan of attack for Yorktown, drawn by Jean Baptiste de Gouvion, October 29, 1781
*National Archives*

**1781 ▶**
French and American engineer officers and sappers and miners played key role in successful siege of Yorktown.

**1783**
Corps of Engineers mustered out of service along with most of the Continental Army.

View of West Point, c. 1834
*Library of Congress*

**1794**
Unified Corps of Artillerists and Engineers established.

**1802 ▶**
Permanent reestablishment of a separate Corps of Engineers and founding of U.S. Military Academy at West Point under Corps supervision.

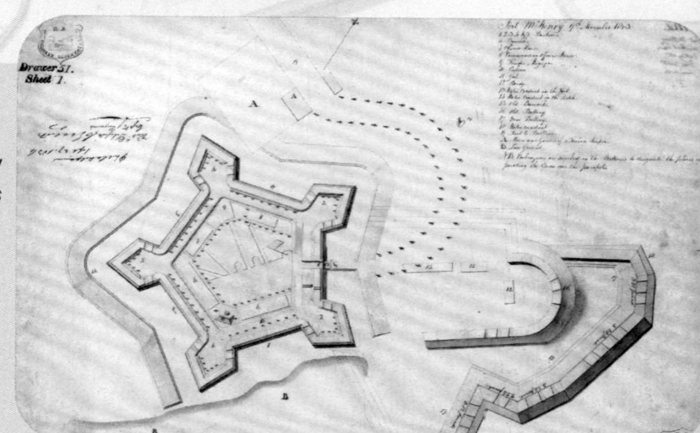


Plan of Fort McHenry
*National Archives*

**1812–1815 ▶**
War of 1812: Coastal harbors heavily fortified by engineers deterred British attacks. Engineer officers first assumed command.

John C. Calhoun
*U.S. Army Collection*

View of the "insulated tablelands" or buttes during Maj. Stephen Long's expedition to the Rocky Mountains, 1820
*Library of Congress*

Nineteenth century survey party in the Sierra Nevada, painting by J.J. Young
*National Archives*

Capt. Henry M. Shreve's snagboat *Heliopolis*

## 1819–1838

**◀ 1819**
Secretary of War John C. Calhoun's report on importance of waterways for national defense and commerce identified role for Army engineers.

**◀ 1819**
Stephen H. Long's expedition of the Missouri River basin pioneered Army engineer involvement in western exploration.

**1824**
An act to improve the navigation of the Ohio and Mississippi rivers initiated permanent civil works construction mission.

**◀ 1824**
General Survey Act authorized use of Army engineers to survey road and canal routes.

**1825**
Corps assumed responsibility for construction and repair of Cumberland Road.

**◀ 1829**
Corps launched first steam-powered snagboat, *Heliopolis,* on the Mississippi River.

**1838**
Creation of separate Corps of Topographical Engineers under Col. John J. Abert.

## 1841–1857

**1841 ▶**
John C. Frémont began a series of western expeditions that ranged to the Rockies and beyond, providing vital information on lands, peoples, and resources of the West.

John C. Frémont's ascent of Snow Peak depicted on a 5-cent stamp
*Smithsonian Institution*

**1846**
Creation of first company of regular U.S. Army engineer troops.

**1846–1848 ▶**
Mexican War: Engineer regulars erected fortifications and joined in assaults while engineer officers performed key reconnaissance missions.

Battle of Monterey, September 23, 1846

**1853 ▶**
Lt. Montgomery C. Meigs began work on a water supply system, the Washington Aqueduct, which still supplies water for the Nation's capital and is still operated by the Corps of Engineers.

Cabin John Bridge, shown here in 1863, built to carry water from the Potomac River over Cabin John Creek to the Washington water supply system

**1853–1858**
Pacific Railroad surveys involved Topographical Engineers in exploration and documentation of the West.

**1857 ▶**
Lt. Gouverneur K. Warren completed his map of the northern plains, the most detailed and accurate to date.

Gouverneur K. Warren as a Major General

# 1861–1863

Henry L. Abbot as a general officer

### ◀ 1861

Humphreys-Abbot *Report Upon the Physics and Hydraulics of the Mississippi River* won the respect of engineers around the world and decidedly influenced the development of river engineering in America.

Company A, Battalion of U.S. Engineers, 1865

### ◀ 1861–1865

Civil War: A battalion of regular U.S. Army engineer troops, with various volunteer engineer and pioneer units, cleared obstacles, constructed roads and bridges, laid down ponton bridges, and erected field fortifications. Several engineer officers commanded combined troops, while others conducted reconnaissance and directed siege operations.

Capitol dome under construction, 1861

### ◀ 1863

New Capitol dome completed under supervision of engineer officer Montgomery C. Meigs.

### 1863

U.S. Army engineers constructed 2,200-foot ponton bridge over the James River, one of the longest ponton bridges in the history of warfare.

Seal of the unified Corps of Engineers

### ◀ 1863

Corps of Engineers and Corps of Topographical Engineers reunified.

## 1866–1883

**1866 ▶**
Engineer School of Application founded at Willets Point, N.Y. Chief of Engineers' role as Inspector of West Point ended as superintendency of the Academy opened to all branches of the U.S. Army.

**1867 ▶**
Control of District of Columbia public parks and monuments given to the Office of Public Buildings and Grounds under the Chief of Engineers until 1933.

**1875 ▶**
Captain William Ludlow's expedition to Yellowstone identified a critical need to protect and improve the park.

**1878**
Three-person commission, including by law an engineer officer, replaced elected government in the District of Columbia until 1967.

**1879**
Mississippi River Commission created to execute a comprehensive flood control and navigation plan on the Lower Mississippi.

**1882 ▶**
In first authorized emergency operation, Corps used its vessels to deliver relief supplies to flood victims.

**1883**
Congress designated Corps to make improvements in Yellowstone Park.

Officers mess, Willets Point, N.Y.

Brig. Gen. Montgomery Meigs managed many Corps construction projects in the District of Columbia from the 1850s to the 1880s.

*National Archives*

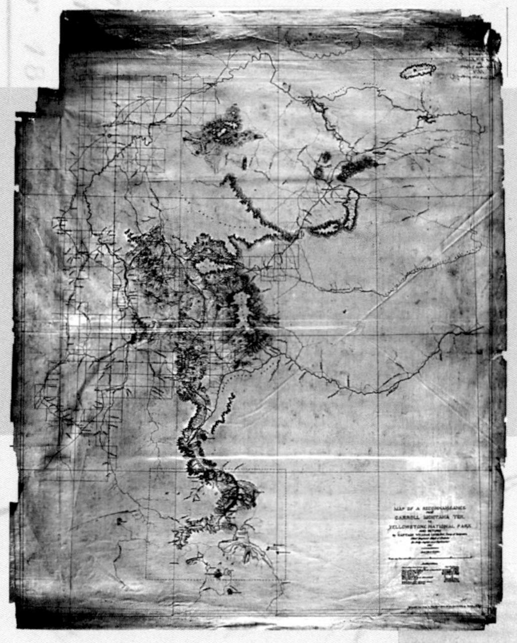

Captain Ludlow's map of reconnaissance from Carroll, Mont., to the site of Yellowstone National Park

U.S. steamer *Montana* at St. Paul, Minn.

Washington Monument nearing completion, early 1884
*National Archives*

The Library of Congress, c. 1897
*National Archives*

Military railroad in the Philippines

The Engineer School academic building, on present-day Ft. McNair, Washington, D.C.

## 1884–1902

**◀ 1884**
Construction of Washington Monument completed by Army engineers.

**1884**
First Corps reservoirs completed at Winnibigoshish, Leech Lake, and Pokegama, Minn.

**1885**
Davis Island Lock and Dam, just south of Pittsburgh, completed—the largest Chanoine wicket dam in the world.

**1888**
Chief of Engineers created five engineer divisions based on geographical regions.

**◀ 1897**
Library of Congress building completed.

**◀ 1898**
Spanish-American War: U.S. Army engineers erected landing piers, built bridges and roads, and repaired and operated railroads in Cuba, Puerto Rico, and the Philippines.

**1899**
Refuse Act gave Corps authority to regulate obstructions to navigation.

**◀ 1901**
Engineer School moved from Willets Point to Washington Barracks, Washington, D.C.

**1902**
Board of Engineers for Rivers and Harbors established to examine costs, benefits, and the need to improve waterways. The board was disestablished in 1993.

# 1911–1927

**1911**
Using a cofferdam, Corps raised wreck of the battleship *Maine* in Havana Harbor.

**1914 ▶**
Panama Canal completed under supervision of U.S. Army engineer officers. Engineer officers served as governors of the Canal Zone from 1914 to 1979.

Steam shovel at work in the Culebra Cut; a lithograph in a series on the Panama Canal by Joseph Pennell, noted American artist and illustrator
*USACE Museum Collection*

**1917 ▶**
Congress passed first federal Flood Control Act.

During a 1912 flood, residents of Hickman, Ky., find refuge on levees and rooftops.

**1917–1918 ▶**
World War I: U.S. Army engineers served in combat; built ports, roads and railroads; organized first U.S. Army tank units; and developed chemical warfare munitions.

World War I Army engineers building a corduroy road

**1919**
Engineer School moved to Camp A.A. Humphreys, Va. (later renamed Fort Belvoir).

**1925 ▶**
Wilson Dam completed with major hydroelectric power component at Muscle Shoals on the Tennessee River.

The power generators at Wilson Dam, Tenn., under construction, 1926

**1927**
Congress authorized 308 Reports to present plans for multipurpose improvement of navigable streams.

## 1927–1939

Sandbagging a levee during flooding of the Mississippi River

### ◀ 1927

Flood devastated Mississippi River Valley and demonstrated insufficiency of "levees only" policy.

Maj. Gen. Edgar Jadwin, Chief of Engineers, 1926–1929

### ◀ 1928

Jadwin Plan becomes basis for landmark Flood Control Act that adopts a comprehensive plan for flood control on the Lower Mississippi River. Plan includes the use of floodways and spillways in addition to levees.

Dashields Locks and Dam on the Ohio River near Glenwillard, Pa., opened for navigation August 1929.

### ◀ 1929

Nine-foot channel completed on the Ohio River.

President Roosevelt visited Ft. Peck during the New Deal Era.

### ◀ 1933

During the Roosevelt administration, Corps' New Deal public works program included Fort Peck, Bonneville, Conchas, and Tygart dams.

### 1936

Flood Control Act made flood control a federal policy and officially recognized the Corps as the major federal flood control agency.

### 1939

Nine-foot navigation channel completed on the Upper Mississippi.

## 1940–1945

**1940 ▶**
Corps took over airfield construction from the Quartermaster Corps' Construction Division.

Hangar at Lowry Field, Colo., 1940

**1941**
Congress transferred Army construction and real estate programs to the Corps of Engineers.

**1942**
Manhattan Engineer District established to oversee construction of production facilities for the atomic bomb.

**1942 ▶**
Engineers completed a 1,500-mile pioneer road, called the Alaska or ALCAN Highway, between Dawson Creek, British Columbia, and Fairbanks, Alaska.

Meeting of bulldozers at Beaver Creek, Yukon Territory, along the ALCAN Highway in 1942

**1943 ▶**
Construction of the Pentagon completed fifteen months after groundbreaking.

An aerial view of the completed Pentagon

**1944**
Flood Control Act authorized Corps to develop recreational facilities on Corps' projects and to develop water projects in the Missouri River Valley in accordance with the Pick-Sloan Plan.

**1945 ▶**
Construction, begun in late 1942, completed on Ledo Road, stretching through some of the world's most difficult terrain from India to the old Burma Road near the Chinese border.

Maj. Gen. Lewis A. Pick in the first convoy on the Ledo Road, May 20, 1945

The dredge *Poseidon* clearing the Corinth Canal in Greece, 1947

Engineers prepare a bridge for demolition in Korea
*National Archives*

Nuclear power plant at Ft. Belvoir, Va.

Nike Ajax missile battery

# 1946–1958

## 1946
Corps began hospital construction program for the Veterans Administration.

## ◀ 1946–1949
Corps' Grecian District supervised postwar reconstruction to restore damaged Greek transportation and communication network to check communist expansion.

## ◀ 1950–1953
Korean Conflict: Engineers destroyed bridges and mined roads to obstruct the enemy, and built bridges and roads to assist advance of American forces. Engineers frequently fought as infantry.

## 1950s
Corps built early warning facilities and air bases in Greenland, Morocco, and Libya.

## ◀ 1952
Corps assigned responsibility for the Army Nuclear Power Program.

## ◀ 1954
Construction of first Nike Ajax missile battery completed.

## 1958
Corps completed work on the American portion of the St. Lawrence Seaway.

## 1960–1976

**1960 ▶**
Corps of Engineers Ballistic Missile Construction Office established to build launch sites and related facilities for intercontinental ballistic missiles.

**1961**
Foreign Assistance Act initiated Corps involvement in reimbursable programs through the State Department's Agency for International Development.

**1961 ▶**
Corps began construction for National Aeronautics and Space Administration (NASA), including the Manned Spacecraft Center in Houston, Texas, and John F. Kennedy Space Center in Fla.

**1962**
In U.S. Army reorganization, Corps lost control of Engineer School and engineer troops but retained responsibility for engineering, construction, and real estate services required by the Army, Air Force, and NASA.

**1963–1973**
War in Vietnam: Forty thousand Army Engineers support combat operations in Southeast Asia.

**1967 ▶**
Rome plow introduced to enhance engineer jungle-clearing operations during Vietnam War.

**1970**
National Environmental Policy Act, signed on January 1, established requirement for environmental impact statements.

**1971–1976**
Corps constructed bulk-mail handling centers for the U.S. Postal Service.

Titan I ICBM in firing position

Vehicle Assembly Building at Cape Kennedy

Depiction of a Rome Plow clearing jungle

Assistant Secretary of the Army (Civil Works) Victor V. Veysey

The Corps' Distinctive Unit Insignia

Tennessee-Tombigbee Waterway, Miss.

## 1972–1985

### 1972
Clean Water Act of 1972 Amendments authorized Corps to regulate dredging and dumping activities in U.S. wetlands.

### ◀ 1975
First Assistant Secretary of the Army for Civil Works named to position originally created in 1970 legislation.

### 1975
Corps redesignated as a combat arms branch.

### 1976
Middle East Division established in Riyadh as Saudi Arabia construction program expanded. Division disestablished in 1986.

### ◀ 1979
Corps of Engineers became an Army Major Command (MACOM).

### 1982
Design and construction effort begun in support of Environmental Protection Agency's Superfund cleanup program.

### 1982
Israeli air bases completed in program initiated in 1979 by Camp David Accords.

### 1983
Defense Environmental Restoration Program enlarged the Corps' environmental work relating to military installations.

### ◀ 1985
Tennessee-Tombigbee Waterway, largest navigation project in Corps' history, completed 13 years after construction began in 1972.

# 1986–1999

## 1986 ▶

The Water Resources Development Act of 1986 brought major change in financing by requiring nonfederal contributions toward most federal water resources projects.

## 1988

The Engineer School relocated to Fort Leonard Wood, Mo.

## 1990–1991

Desert Shield/Desert Storm: Corps provided construction and real estate support.

## 1991

Recovery effort in Kuwait initiated through the Kuwait Emergency Recovery Office.

## 1991

Beginning of successive rounds of base closures under a presidentially appointed realignment commission.

## 1992 ▶

Corps undertook major disaster recovery in wake of hurricanes Andrew and Iniki.

## 1993

Assistant Chief of Staff for Installation Management created on the Department of the Army staff. The new office absorbed many of the functions of the Assistant Chief of Engineers.

## 1996

Groundbreaking ceremony for the Olmsted Lock, the last major lock modernization project on the Ohio River in a program begun in the 1950s.

## 1997

Formerly Used Sites Remedial Action Program transferred from Department of Energy to the Corps of Engineers.

## 1999 ▶

Dedication of the Seven Oaks Dam of the Santa Ana River Mainstem Flood Control Project—potentially the Corps' last big dam project.

President Reagan signs the Water Resources Development Act, 1986.

Installing temporary roofs following Hurricane Andrew

Seven Oaks Dam, Ca.

*Photo by Dave Schumaker*

Ruins of the World Trade Center, New York City, September 2001

Personnel of the Gulf Region Division in Baghdad, Iraq

Personnel from the Gulf Region Division discuss construction with Iraqi contractors in Sulaymaniyah, Iraq

Chinook helicopter prepares to lift 5,000-pound sandbags to repair damaged flood walls.

## 2000–2005

### 2000
Congress approved the Comprehensive Everglades Restoration Plan with the Corps designated as the lead agency.

### 2001
The Upper Mississippi River/Illinois Waterway Navigation Study and its recommendation for the construction of new and larger locks generated substantial controversy and opposition.

### ◀ 2001
9/11: Corps of Engineers responded to terrorist attacks at the World Trade Center and the Pentagon.

### 2001
Missouri River Master Water Control Manual became increasingly controversial because of environmental issues and competing interests in the river basin.

### 2002
After the fall of the Taliban regime in Afghanistan, the Corps of Engineers began a program to construct facilities for the Afghan National Army.

### ◀ 2003
Soon after coalition forces entered Iraq, the Corps began to restore the Iraqi oil and electrical infrastructure.

### ◀ 2004
The Gulf Region Division established in Baghdad to manage the reconstruction program.

### ◀ 2005
Hurricanes Katrina and Rita ravaged the Gulf Coast and subsequent storm surges overwhelmed the protective levees around New Orleans, flooding the city.

# 1 The Making of the Corps

Battle of Bunker Hill, June 17, 1775, *by H. Charles McBarron*
*U.S. Army Center of Military History*

# The Revolutionary War

When Congress organized the Continental Army on June 16, 1775, it provided for a Chief Engineer and two assistants with the Grand Army and a Chief Engineer and two assistants in a separate department, should one be established. Colonel Richard Gridley of Massachusetts was an artillerist in that colony's militia and a veteran of decades of colonial warfare against the French, and thus one of the few patriots with experience in the design and construction of batteries and fortifications. Gridley became General George Washington's first Chief Engineer. Another native of Massachusetts, Rufus Putnam, who succeeded Gridley as Chief Engineer in 1776, was one of his assistants while the Army remained in Boston.

From the start, the predominantly defensive nature of the war convinced Washington he would need even more trained engineers, but he was continually frustrated in his efforts to find them. Qualified engineers were scarce because formal schooling in siegecraft, the erection of field fortifications, and technical

*Society of American Military Engineers plaque honoring Col. Richard Gridley's actions at Breed's Hill*

*Washington takes command of the Continental Army at Cambridge, Mass., July 1775*

*National Archives*

*Louis Lebègue Duportail, Chief Engineer, 1777–1783 by Charles Willson Peale*

*Independence National Historical Park Collection*

subjects was practically nonexistent in America at the time. In response to Washington's plea for more engineers, Congress turned to France, which was an enemy of Britain and the center of technical education in Europe. The French also had a long tradition of military engineering.

Beginning in 1776, Frenchmen began to arrive in America to serve as engineers. Before the end of 1777, Congress had promoted one of them, Louis Duportail, to brigadier general and Chief Engineer, a position he held for the duration of the war in spite of a period of capture and imprisonment by the British following the Battle of Charleston. Frenchmen, joined by other foreigners, dominated the ranks of the engineers throughout the war.

When Duportail took command of the engineers, he renewed the pressure begun by his predecessor to establish a permanent, separate, and distinct engineering branch of the Army. His proposal included a provision for companies of engineer troops, to be known as Sappers and Miners, with American officers. From their ranks would come the engineer officers to replace the French when they returned home.

On May 27, 1778, Congress finally authorized three companies of Sappers and Miners who were to receive instruction in erecting field works—the first step in technical education—and were to direct fatigue parties, repair damaged works, and erect new ones. Recruitment continued for more than two years, and the three companies were not activated until August 2, 1780. Meanwhile, on March 11, 1779, Congress passed a resolution that formed the engineers in the Continental Army into the Corps of Engineers that Duportail had sought.

Despite the shortage of engineers and the delay in forming companies of engineer troops, the Army's engineers made numerous contributions to the war. Engineer officers reconnoitered enemy positions and probable battlefields, wrote useful reports based on their observations, oversaw the construction of fortifications, and drew detailed maps for commanders. Congress relieved some of the mapping burden when it appointed Robert Erskine as

*Boston Area Fortifications drawn by John Trumbull depicting American positions surrounding Boston in September 1775*

*John Trumbull,* Autobiography, *1841*

Geographer of the Army in 1777. Erskine and his successor, Simeon DeWitt, employed several assistants, as did Thomas Hutchins, whom Congress appointed as Geographer for the Southern Army in 1780. Following this precedent, Congress would add Topographical Engineers to the Corps of Engineers in 1813 and create a Topographical Bureau in the Engineer Department in 1818.

Engineer officers often took action that helped achieve decisive results on the battlefield. One such incident occurred during the Siege of Boston. In February 1776, General Washington's council of war decided to draw the British out of Boston by erecting works on the unfortified Dorchester Heights. To achieve surprise, the Army needed to move quickly, but the ground was frozen more than a foot deep. Colonel Rufus Putnam, Washington's Chief Engineer at the time, offered an innovative solution to the problem. He recommended using chandeliers—wooden frames filled with bundles of sticks—to raise walls above ground. To the astonishment of the enemy, the Continentals erected the chandeliers overnight on March 4. When the British determined three days later that Dorchester Heights could not be taken, they found that their hold on Boston was no longer tenable and evacuated the city.

The next year, Lieutenant Colonel Thaddeus Kosciuszko, a native of Poland commissioned as an engineer officer in the Continental Army, placed obstructions that significantly impeded Burgoyne's advance toward Albany after the fall of Fort Ticonderoga. Later, Kosciuszko helped design the network of defenses at West Point, and in 1781 he was instrumental in enabling Nathaniel Greene's Southern Army to evade capture by the enemy.

During the difficult winter of 1777–1778, Washington followed Duportail's admonition to avoid general battle and instead wear down the British at Philadelphia while avoiding attack. "We should not forget that in war, to advance or retreat are neither honorable nor dishonorable; that

*Thaddeus Kosciuszko by Charles Willson Peale*

*Independence National Historical Park Collection*

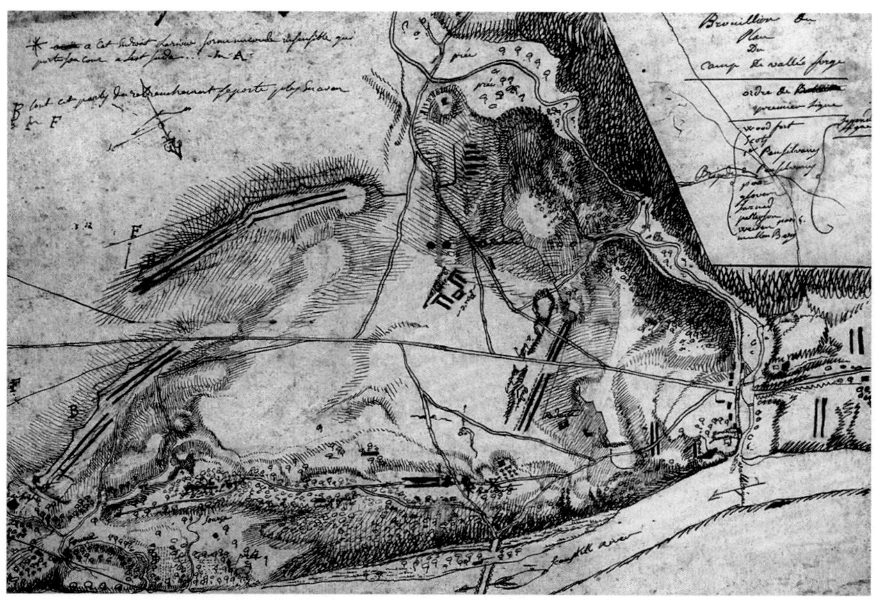

*The Camp at Valley Forge. A sketch in Duportail's hand showing the entrenchments he planned.*

*Historical Society of Pennsylvania*

it is [at] the end of a Campaign that the Prize is given, and that Glory is his reward who has gained his end," Duportail noted in recommending that Washington keep his forces at Valley Forge. This strategy helped preserve the Army and compelled the British to evacuate Philadelphia the next summer.

The Corps of Engineers and its companies of Sappers and Miners enjoyed their finest hour in October 1781 at Yorktown, where Washington conducted a siege in the classical manner of Sebastien de Vauban, the great French master of siegecraft. Engineer officers, numbering thirteen in the combined French and American armies, performed crucial reconnaissance and, with the fifty men in the Sappers and Miners, planned and executed field works. In addition, the Sappers and Miners assembled fortification materials, erected gun platforms, transported cannon and ammunition, and cleared the way for the decisive infantry assault on Redoubt 10. After the battle, Washington cited Duportail for conduct that afforded "brilliant proofs of his military genius, and set the seal of his reputation."

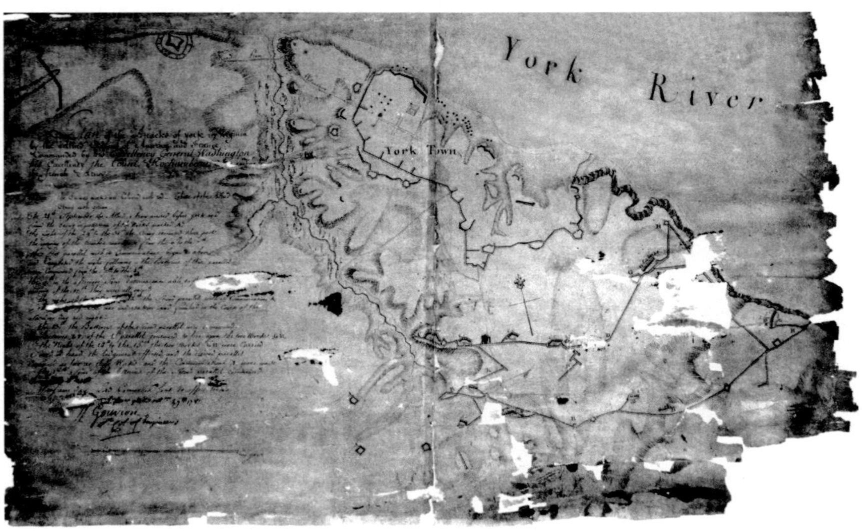


*British defenses and the American and French siege works at Yorktown drawn by Col. Gouvion of the Corps of Engineers, October 29, 1781*
*National Archives*

When the Revolutionary War ended in 1783, a debate followed on the peacetime nature of the Army. Proposals regarding the engineers varied. They included merging the engineers with the artillerists and establishing an academy to provide training. Those who favored a centralized system of fortifications, which would need engineers to build and maintain them, believed that retaining an engineer presence in the Army was necessary. Two arguments in favor of retaining the engineers drew directly upon Revolutionary War experience. Without a permanent, trained Corps of Engineers, it was maintained, the new Nation would be forced to call on foreigners again in time of war. Moreover, as the Revolutionary War had demonstrated, it was extremely difficult to put together an effective technical organization in a short time. But Congress did not approve a peacetime Army, and with that decision went any hope of retaining the Corps of Engineers. By the end of 1783, the Corps and its companies of Sappers and Miners had mustered out of service.

# Congress Recognized a Revolutionary War Engineer as a Hero for His Role in Taking Stony Point in July 1779

On June 1, 1779, the British captured Stony Point, New York, on the western side of the Hudson River, and Verplanck's Point directly across the river to the east. Possession of the forts brought a key part of the river under enemy control and threatened the American position less than fifteen miles to the north at West Point. After reinforcing Stony Point, the British commander regarded it as a "little Gibraltar."

Recognizing the danger, General George Washington planned a daring surprise assault. On the night of July 15–16, Lieutenant Colonel François de Fleury, an engineer in command of a battalion in the 1st Regiment of Brigadier General Anthony Wayne's Corps of Light Infantry, led one of two simultaneous attacks on Stony Point. In the hour after midnight, the twenty-nine-year-old de Fleury single-handedly struck the colors of the British 17th Regiment of Foot. Invaders and defenders engaged in furious hand-to-hand combat. The whole encounter was brief. At 2 a.m., Wayne triumphantly wrote Washington, "The fort and garrison…are ours. Our officers and men behaved like men who are determined to be free."

A few days later, Washington abandoned the fort for lack of resources. The British quickly reoccupied the site, temporarily making it stronger than ever. But reinforcements never arrived, so the Redcoats gave up the position for good in October. Stony Point was a timely boost to American morale. It was, according to one historian, "a successful attack upon British regulars in a fortified position, with the bayonet alone," resulting in "an achievement unparalleled up to that time." It also marked the last major battle of the war in the north.

*Storming of Stony Point depicted in a nineteenth century lithograph*

In recognition of the bold, decisive action at Stony Point, Congress awarded a gold medal to Wayne and silver medals to de Fleury and Major John Stewart, who commanded a battalion in the 2d Regiment of Wayne's corps. Congress noted that de Fleury and Stewart "exhibited a bright example to their brother soldiers, and merit in a particular manner the approbation and acknowledgment of the United States." De Fleury, one of several French engineers to volunteer for service in the Continental Army, was the only foreigner so honored during the Revolutionary War.

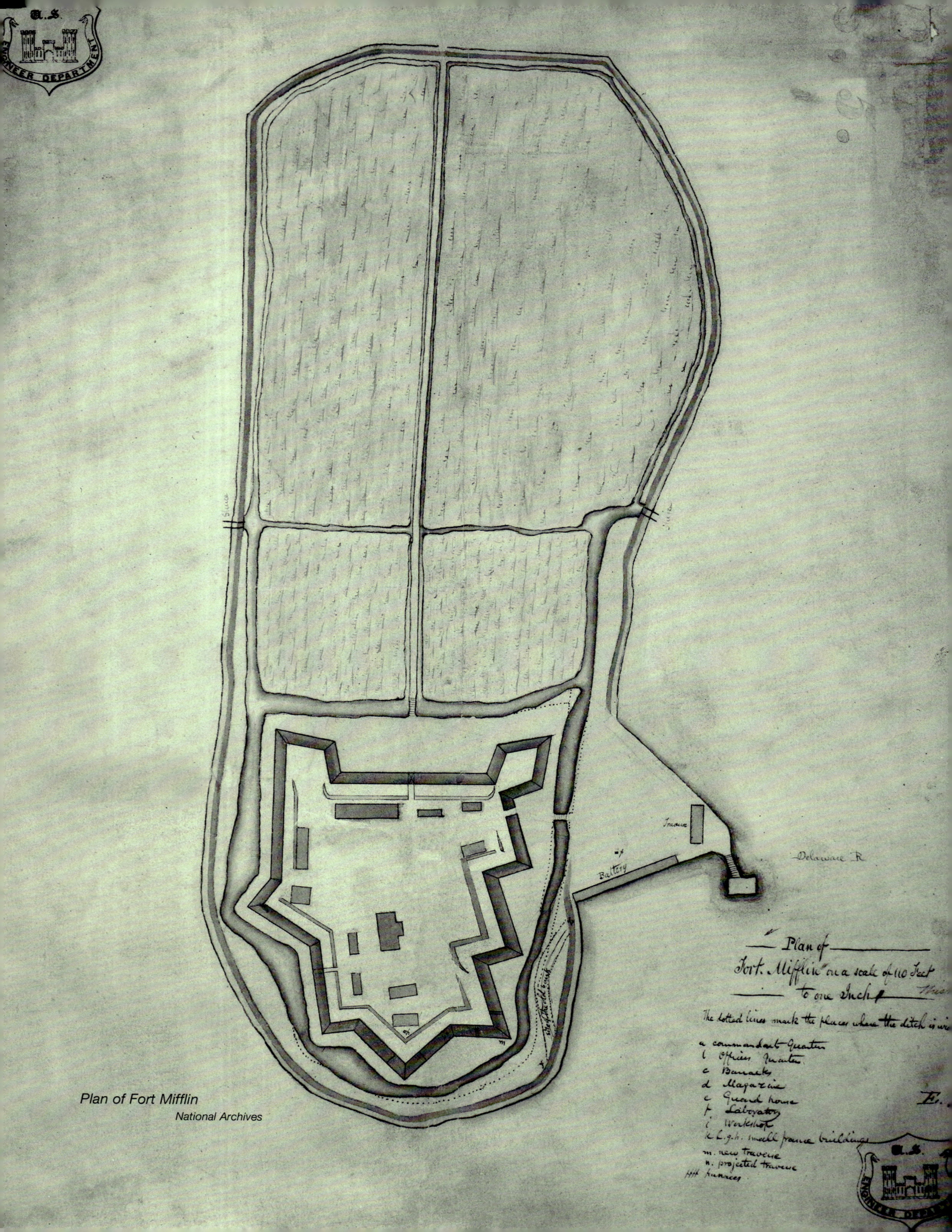


Plan of Fort Mifflin

National Archives

# Artillerists and Engineers, 1783–1802

After the American Revolution, there was strong opposition to the establishment of a large, permanent, national army. Indeed, at one point in the summer of 1784, the surviving U.S. military establishment consisted solely of an infantry regiment and a company of artillery stationed at West Point, New York; however, Congress soon approved the formation of an additional line unit, the 1st American Regiment, to construct forts and protect surveying parties on the new western frontier.

When a strengthened federal government under the new U.S. Constitution was launched in 1789, Secretary of War Henry Knox recommended "a small corps of well-disciplined and well-informed artillerists and engineers." Nevertheless, no engineers served in the U.S. Army until March 1794, when war with Britain threatened. Suddenly there was an acute need to upgrade neglected coastal fortifications and construct new ones. At that time, Congress authorized President Washington to appoint temporary engineers to direct the fortification of

*Portrait of Henry Knox by James Harvey Young, 1873*

*U.S. Army Center of Military History*

key harbors. Among those named were Major Pierre L'Enfant and Major Stephen Rochefontaine, veterans of the Revolutionary War Corps of Engineers.

Seizing the opportunity, Knox again urged Congress to approve the plan he and others such as L'Enfant and Duportail had earlier advanced. A corps combining artillerists and engineers, he argued, would provide the additional trained troops now needed to garrison the coastal fortifications. The new corps was to be commanded by a lieutenant colonel

*Only known authentic likeness of Pierre Charles L'Enfant. Silhouette by Sara DeHart, c. 1785.*
*Diplomatic Reception Room, U.S. Department of State*

and to have four battalions, each commanded by a major and consisting of four companies.

Heeding this advice, on May 9, 1794, Congress established a single Corps of Artillerists and Engineers, consisting of one regiment. This action returned engineers to the ranks of the Army for the first time in more than ten years and ensured that an engineering presence, established with the appointment of the Army's first Chief Engineer in 1775, would continue in the new U.S. Army.

Although international tensions eased in the latter half of 1794 and jeopardized the whole effort, Congress resolved to continue a seacoast defense program. By the end of the year, there were single-company garrisons of artillerists and engineers at Fort Jay (New York); Fort Mifflin (Philadelphia); Fort Whetstone, later McHenry (Baltimore); and Fort Johnson (Charleston). The following February, Rochefontaine was commissioned a lieutenant colonel and took command of the Corps. At the same time, a school to train U.S. Army officers took shape at West Point, New York.

In 1798, war with France appeared likely, so Congress added a second regiment to the Corps. By the time Thomas Jefferson became president in 1801, it had become clear that the united Corps was not producing the desired well-educated, scientific body of engineer officers. In 1802, a thrifty Congress again reduced the military establishment and separated the artillerists and engineers. The united corps, which so many Revolutionary War engineers had supported, was thus short-lived; however, an Army engineering branch would emerge from the peacetime reduction. On March 16, 1802, Congress permanently established a separate U.S. Army Corps of Engineers and the U.S. Military Academy at West Point as the Nation's first engineering school.

*Parade Field at West Point, c. 1790*
*U.S. Military Academy Library*

# French Engineers Defended Early Capital City

*Oil painting of Fort Mifflin c. 1872 by Brig. Gen. Seth Eastman*

*Architect of the Capitol Collection*

At the end of the Revolutionary War, General George Washington recommended retaining a regular, standing force to garrison forts and one or more academies to provide Americans with engineering and military training. Otherwise he predicted that domestic security in the future would depend entirely on the assistance of foreigners.

Congress failed to act on his recommendation. As president in 1794 he faced the prospect of renewed war with Britain. Coastal forts lay in disrepair, and America's tiny Army lacked artillerists and engineers. Congress quickly appropriated funds to fortify nearly twenty ports and harbors and created a combined Corps of Artillerists and Engineers.

Few native-born Americans were available to plan and oversee the required defenses, but Secretary of War Henry Knox knew that some one dozen Frenchmen, who had either served in the French or Continental armies during the Revolution, were in the United States. Most had returned to France after the Revolution but fled in the wake of the French Revolution's Reign of Terror. They found their way back to the United States either directly or by way of Santo Domingo. Knox employed seven of these Frenchmen as "temporary engineers" without military rank to supervise the new work. Each was assigned a section of the Atlantic coastline.

Greatest attention and funding focused on Philadelphia, the Nation's temporary capital (1790–1800). Initially Knox assigned Pierre L'Enfant to oversee improvements at Mud Island (later Fort Mifflin) just below the capital city. After service as an engineer in the Revolutionary War, L'Enfant returned briefly to France in 1783 but took up residence in the United States in 1784. Later he designed the city of Washington.

Convinced that Fort Mifflin provided inadequate protection for a capital city, L'Enfant embarked on an ambitious plan of improvements. He quickly exhausted the available funds and antagonized state officials in the process. Within a year Knox replaced him with another Frenchman and Revolutionary War veteran, Stephen Rochefontaine. More improvements and additional funding followed.

A "quasi" war with France in 1798 led Congress to strengthen the Corps of Artillerists and Engineers and appropriate more funds to defend American shipping, the coastline, and harbors. As Rochefontaine was also commander of the artillerists and engineers, Lewis Tousard, another Frenchman, took over at Fort Mifflin. Once again he got the job because no American possessed the technical qualifications.

Nearly half the funds expended on American forts between 1794 and 1800 went to Fort Mifflin. The experience of having the Nation's defenses planned and executed by foreigners finally convinced Congress to establish a military academy and create a separate Corps of Engineers.

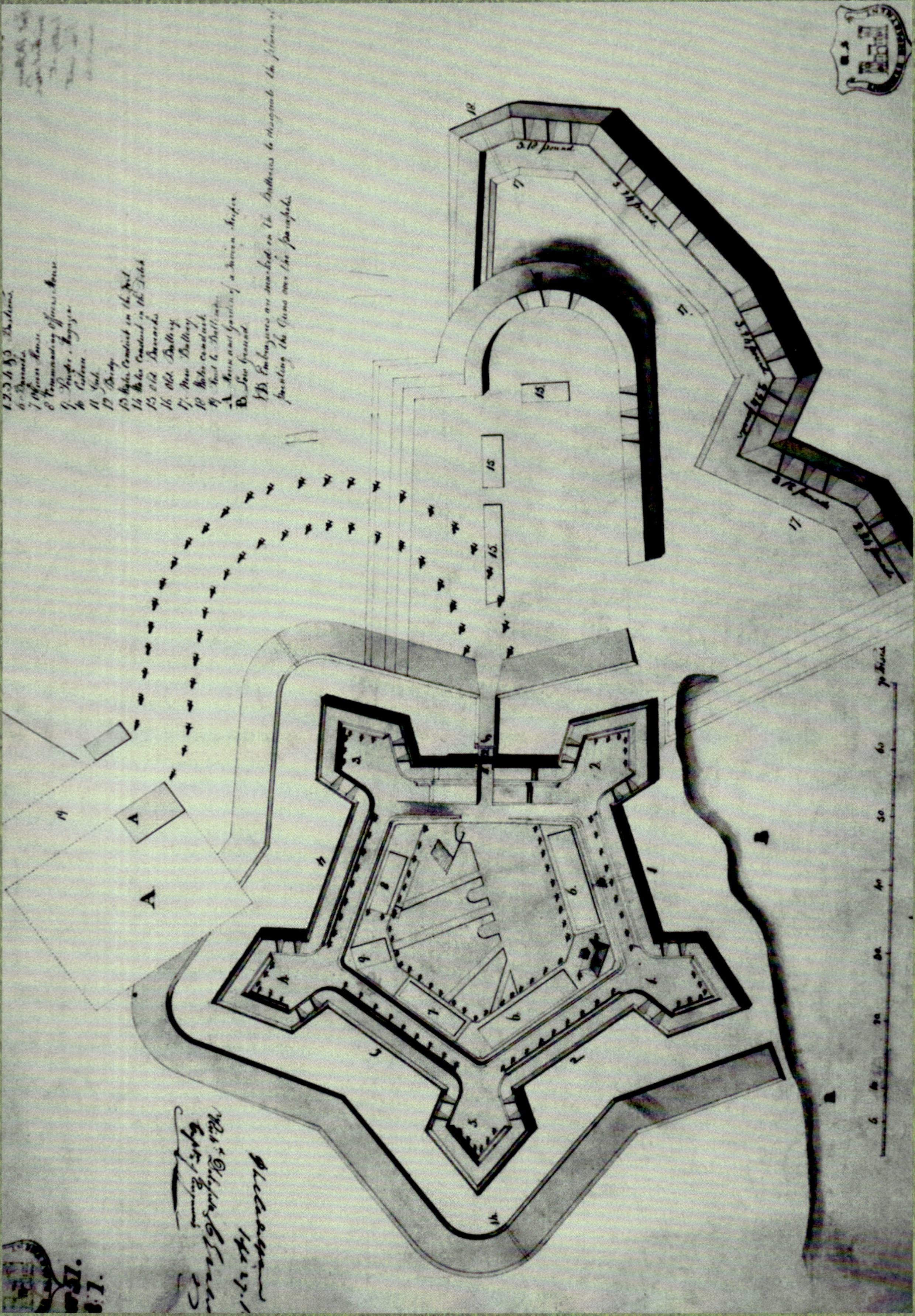

*Plan of Fort McHenry by Capt. Richard Delafield, September 1836*
National Archives

# Engineers in the War of 1812

After the American Revolution, engineer officers did not see formal combat again until the War of 1812. During the years immediately preceding that conflict, engineer officers had worked full time constructing permanent defenses along the Atlantic Coast. The War Department had been debating with the engineers over their desire for command responsibility since 1802. Jonathan Williams, the first superintendent of West Point, had even resigned his position over the issue.

The Corps of Engineers remained small in numbers. When war broke out in June 1812, the Corps' actual strength was only seventeen officers and nineteen enlisted men. (Although Congress had authorized 22 officers and 113 enlisted men for the Corps in April 1812, full strength was not approached until 1815.) West Point graduates dominated the list of engineer officers, and for them the War of 1812 would be their first experience in combat.

During the War of 1812, the record of the Corps was exceptional when compared with the record of the other branches of the U.S. Army, which suffered several notable defeats. Engineer officers assumed command responsibility for the first time. Captain Charles Gratiot, later Chief Engineer, at one point commanded all forces in the Michigan Territory. In 1813, Brigadier General Joseph G. Swift, another future Chief Engineer, commanded line units on Staten Island in addition to Fort Richmond and Hudson Battery. By late the next year, he commanded the entire New York operation, which included more than ten thousand soldiers and civilian volunteers.

*Burning of the U.S. Capitol by the British, August 1814, from mural by Allyn Cox in the House of Representatives Cox Corridor*
*Architect of the Capitol Collection*

*Battle of Lundy's Lane, where U.S. Army engineers figured prominently, July 1814*
*Library of Congress*

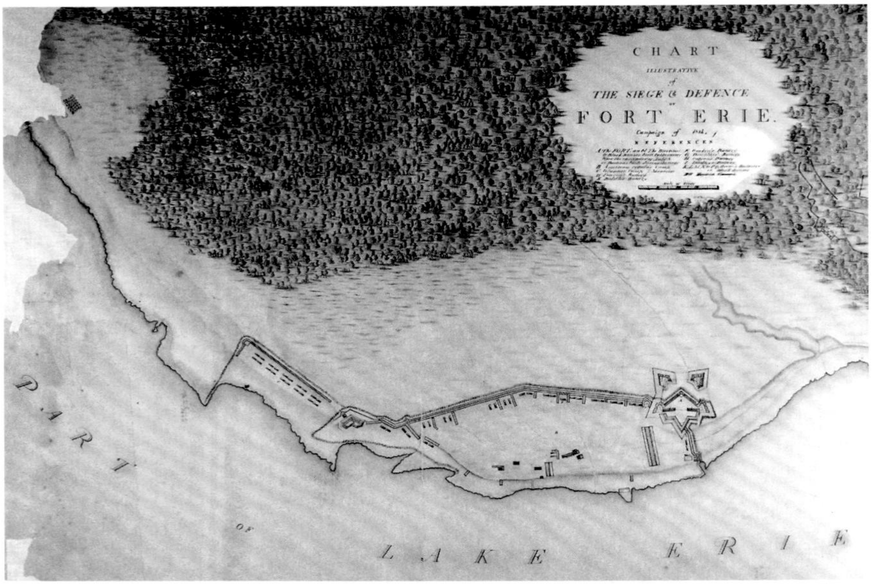


*Map of Fort Erie depicting how Army engineers changed the old British fort into a bastioned work*

*National Archives*

*Bombardment of Fort McHenry by J. Bower*

*Maryland Historical Society*

As the war progressed, the War Department increasingly transferred engineers to serve in the field on the northern frontier. In combat, the engineers performed many of the same tasks they had in the Revolution, including constructing fortifications, reconnoitering and mapping, and assisting the movement of armies. In at least two instances, engineer officers directed construction of quarters.

Still, fortifications were the primary concern of the engineers during the War of 1812, as they had been earlier. Despite the views of later critics, coastal harbors and river towns heavily fortified by the engineers did deter British attack. Notable examples of this were at Fort Meigs in Perrysburg, Ohio, and Fort McHenry in Baltimore, Maryland.

The performance of the U.S. Army engineers in combat between 1812 and 1815 helped them earn respectability and strengthened the military academy at West Point, which had been languishing on the eve of the war. While many battles in this indecisive war ended in a standoff, the results might have been far worse without the contributions of the U.S. Army engineers.

# An Engineer Helped Save New York City from British Attack During the War of 1812

From the beginning of the War of 1812, the British captured American ships, blockaded major ports, and raided towns along the coast. In 1814, British troops even seized Washington, D.C., burning the White House and U.S. Capitol and occupying Alexandria, Virginia. Recalling its own capture by the British during the American Revolution, New York—the Nation's largest city—felt especially threatened.

While British ships cruised just off Sandy Hook, New Yorkers turned to the U.S. Army for help. During most of 1813 and 1814, Brigadier General Joseph G. Swift, Chief Engineer of the Army and superintendent at West Point, directed the city's defenses. Until mid-1814, he concentrated on the harbor's permanent forts.

In the summer of 1814, a reinforced British fleet appeared off New York's coast. Fearing an amphibious attack from the north or east, the city's Committee of Defense asked Swift to take charge of emergency preparations. Quickly, he drew up a plan calling for two lines of field fortifications, one stretching along hilltops outside Brooklyn, the other cutting across Manhattan from the mouth of the Harlem River to the Hudson. Then he began to implement the plan and called upon citizens for support. The response was overwhelming.

Between August and November, thirty-eight thousand people worked on the defenses. Carpenters and pharmacists, brewers and lawyers, butchers and college students, tailors and artists, free blacks and city officials rubbed shoulders in the trenches, wielding axes, shovels, and spades. Organized in parties of 1,200–2,000, often working from sunrise to sundown, and singing to keep their spirits high, they built two lines of field defenses. Volunteers put in a total of more than one hundred thousand workdays. People unable to work contributed money, food, tools—and more than five thousand fascines for the parapets.

Swift oversaw all defense preparations. Before long, he also was plotting strategy; inspecting troops; and directing ordnance, artillery, quartermaster, and medical activities. In the event of a British landing, he intended to lead the main force to repulse them. Impressed by the strength of New York's defenses, the enemy chose easier targets to attack.

In gratitude for Swift's service, the New York Common Council declared him a benefactor of the city, showered him with gifts, and commissioned John Wesley Jarvis to paint his full-length portrait. After the war, to commemorate the Chief Engineer's heroic effort on their city's behalf, officials hung the painting in New York's City Hall.

*Joseph Gardner Swift (1783–1865)*

*Courtesy of the Art Commission of the City of New York*

*Plan of West Point, 1780*

*National Archives*

# The Corps and the Military Academy at West Point, 1802–1866

During the American Revolution, many officers, including General George Washington, the Commander-in-Chief, saw the need for technical education so that the Army would have skilled, native-born American engineer officers in the future. When Congress established the companies of Sappers and Miners in 1778, it stated that the companies were to receive instruction in field works. In subsequent general orders, Washington referred to the Sappers and Miners as "a school of engineering." Regulations issued in 1779 for the Corps of Engineers and companies of Sappers and Miners declared that the Sappers and Miners were to receive instruction at times when they were not exercising duties. The Chief Engineer was to devise an instructional program and appoint engineer officers to give lectures; however, the amount of education actually given the Sappers and Miners during the Revolution was minimal.

During the debate over a peacetime military establishment in 1783, several Army officers proposed establishing an academy at West Point, either as the sole military academy or as one of several academies. Army leaders thought engineers in particular needed formal

*L'Enfant watercolor of West Point, 1780*

*National Archives*

*Dennis Hart Mahan*
*U.S. Military Academy Library*

training. When Congress decided against a peacetime standing Army, the need for an academy disappeared.

Some instruction did occur at West Point from 1794 until 1796, but it was not until March 16, 1802, that Congress reestablished a separate Corps of Engineers to remain at West Point and constitute the U.S. Military Academy. As Chief Engineer, Jonathan Williams, grandnephew of Benjamin Franklin and a man keenly interested in the development of science, became the Academy's first Superintendent. Williams introduced new texts from England and the continent and, by 1808, had broadened the curriculum from its heavy emphasis on mathematics to include engineering. In 1812, Congress created a professorship of engineering at the Academy. It was the first such position at an institution of higher learning in the United States.

Major advances in the organization and the course of study, as well as an honor code and a disciplinary system, followed under Sylvanus Thayer, superintendent from 1817 until 1833. Thayer patterned the reorganization of the Academy on the program he observed at the *Ecole Polytechnique* while on a visit to France. Claudius Crozet, who occupied the professorship of engineering from 1817 to 1823 and who was a graduate of the *Ecole Polytechnique,* introduced numerous French texts in his courses. Later, under professor Dennis Hart Mahan, the Academy's reputation as a school of civil engineering advanced still further. In his lectures, Mahan, an 1824 graduate with a commission in the Corps of Engineers, drew upon his experiences while on duty in Europe (1826–1830). He prepared and added several texts to the West Point curriculum. The most important were *A Treatise on Field Fortification* (1836) and the *Course of Civil Engineering,* which first appeared in 1837.

In 1800, Secretary of War James McHenry emphasized that fortification was only one part of military engineering. The engineer's utility, he declared, "extends to almost every Department of War; besides embracing whatever respects public buildings, roads, bridges, canals and all such works of a civil nature."

*Training in practical military engineering*
*U.S. Military Academy Library*

*A June examination by the Board of Visitors*

U.S. Military Academy Library

*Cadets working with models, Class of 1904*

U.S. Military Academy Library

After the War of 1812, West Point exemplified McHenry's dictum. The Academy was the first school of engineering in America. For many years it produced graduates who, in addition to heroic battlefield achievements, played a major role in the Nation's internal improvement in areas such as mapping, roadbuilding, constructing canals, improving harbors, and building railroads. President Andrew Jackson labeled it "the best school in the world."

The Military Academy continued under the supervision of the U.S. Army Corps of Engineers until 1866, when Congress opened the superintendency to all branches of the Army and placed control of the Academy under the Secretary of War, thus ending the Chief of Engineers' role as Inspector. This change responded, in part, to the fact that the Academy supported the entire Army, not just the engineers; however, mathematics, science, and engineering continued to remain at the center of the curriculum.

# Robert E. Lee as an Army Engineer

Robert E. Lee was a U.S. Army engineer officer from 1829 to 1855. Coming from a well-known family that already had its military heroes, Lee's career as a military officer was virtually foreordained. After preparatory school, Lee, gifted at mathematics, sought admission to West Point.

*Robert E. Lee as a captain*

The number of applicants rejected by West Point far outnumbered those accepted each year so a relative wrote to Secretary of War John Calhoun on Lee's behalf and Lee presented the letter to the Secretary in person to make a positive impression. Family connections to important congressmen further aided him in his quest. On March 11, 1824, Lee received admission to the Academy for the class beginning in the summer of 1825—the delay resulting from a backlog of admitted cadets.

Lee entered the U.S. Military Academy in West Point, New York, on July 1, 1825. An excellent student, he graduated number two in the Class of 1829. As most top graduates did in the nineteenth century, Lee entered the U.S. Army Corps of Engineers as a second lieutenant.

His first assignment was constructing fortifications in Georgia and Virginia, including Fort Monroe. He later supervised navigation work along the Mississippi River, and for five years he oversaw upkeep of the forts in New York Harbor. In 1846, Lee was assigned to the campaign in northern Mexico. He eventually participated in all the main battles from Vera Cruz to Mexico City, and received a final brevet to colonel for his valor at Chapultepec. From 1848 to 1852, Lee was the supervising engineer for construction of Fort Carroll near Baltimore, Maryland.

In 1852, Lee accepted an assignment as Superintendent at the United States Military Academy, a position reserved for Corps of Engineer officers until 1866. While heading the Academy, he instituted many important changes to the curriculum. He also encountered cadets who would be prominent in the coming Civil War—including Union generals James McPherson, Philip Sheridan, and O. O. Howard, and Confederate generals John B. Hood and Jeb Stuart. Lee's son, G. W. Custis Lee, also served as a cadet while his father was Superintendent.

Lee left the Corps of Engineers in 1855 when he accepted an assignment as a lieutenant colonel in the 2d Cavalry Regiment. In 1859, he led a contingent of Marines to retake the armory seized by John Brown at Harper's Ferry. Offered command of all Union forces at the outset of the Civil War, Lee chose loyalty to his state and the South and accepted a Confederate generalship.

Douglas Southall Freeman, Lee's biographer, said that Lee's mind was mathematical and his imagination that of an engineer, and that his training as an engineer worked to his advantage when strategizing. Lee recognized and admired the engineers among his opponents, many of whom were his former students or fellow engineer officers. Asked to name the best Union general, Lee answered George B. McClellan, commander of the Army of the Potomac and Corps of Engineers officer from the West Point Class of 1846.

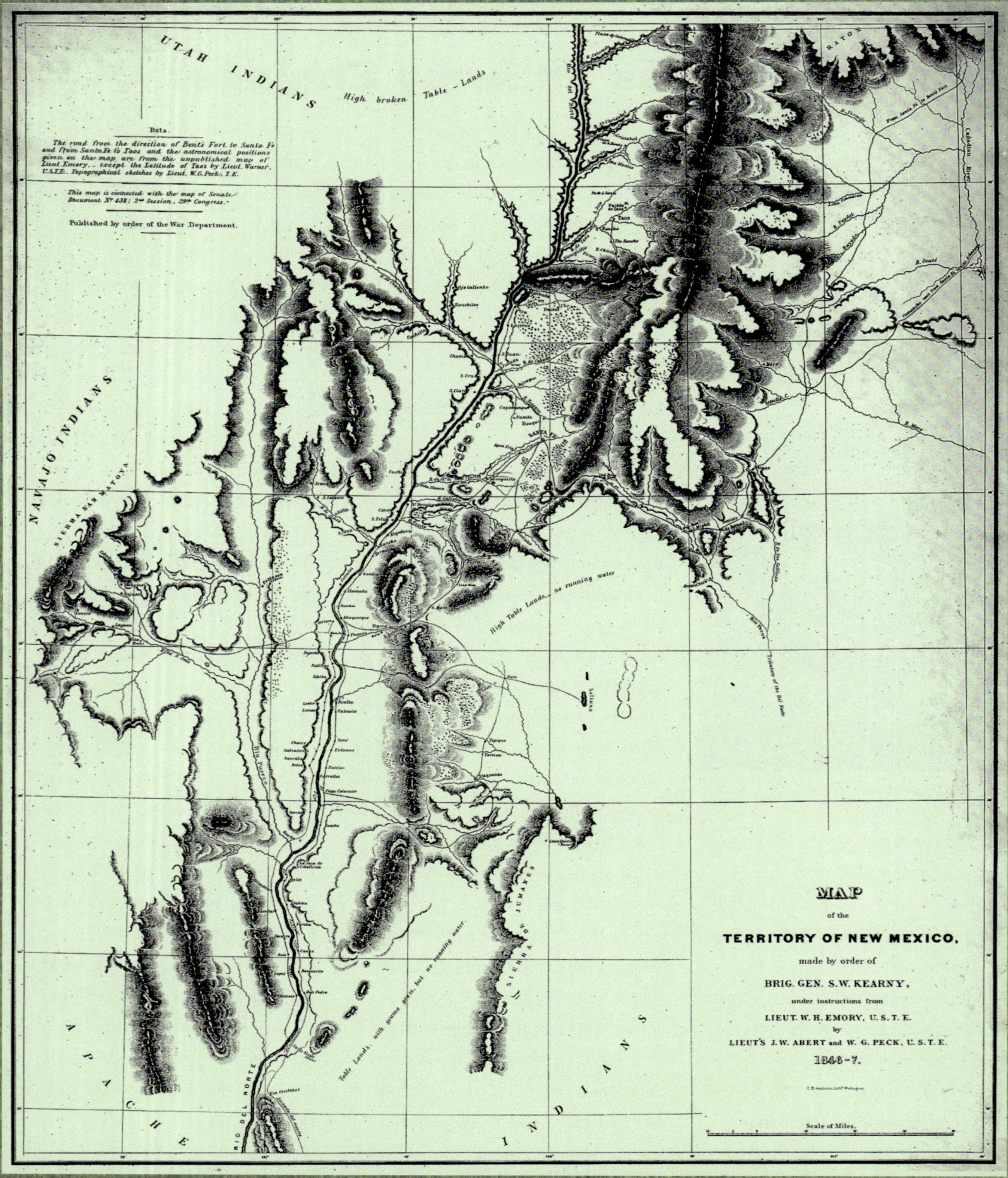


*Map of the Rio Grande Valley, drawn in 1846–47 for Mexican War reconnaissance*

*National Archives*

# The Topographical Engineers

The U.S. Army played a key role as the young Nation rapidly expanded during the nineteenth century. During his first inaugural address in 1801, President Thomas Jefferson said, "However our present interests may restrain us within our own limits, it is impossible not to look forward to distant times, when our rapid multiplication will expand itself beyond those limits and cover the whole...continent." Seizing upon an opportunity to greatly increase the land size of the United States, Jefferson negotiated with Napoleanic France for the Louisiana Purchase. Soon thereafter, the imaginative president sought to have this large expanse explored, with the ultimate goal of finding a Northwest Passage. The reconnaissance of the Trans-Mississippi West began with the four-thousand-mile epic journey of Lewis and Clark in 1804–1806. They traveled the length of the Missouri, Clearwater, Columbia, and Snake rivers to the Pacific Ocean.

Another ten years would pass before the government began to professionalize official exploration. In 1816, topographical officers, known

*View of "insulated tablelands" or buttes during Maj. Stephen Long's expedition to the Rocky Mountains, 1820*

*Library of Congress*

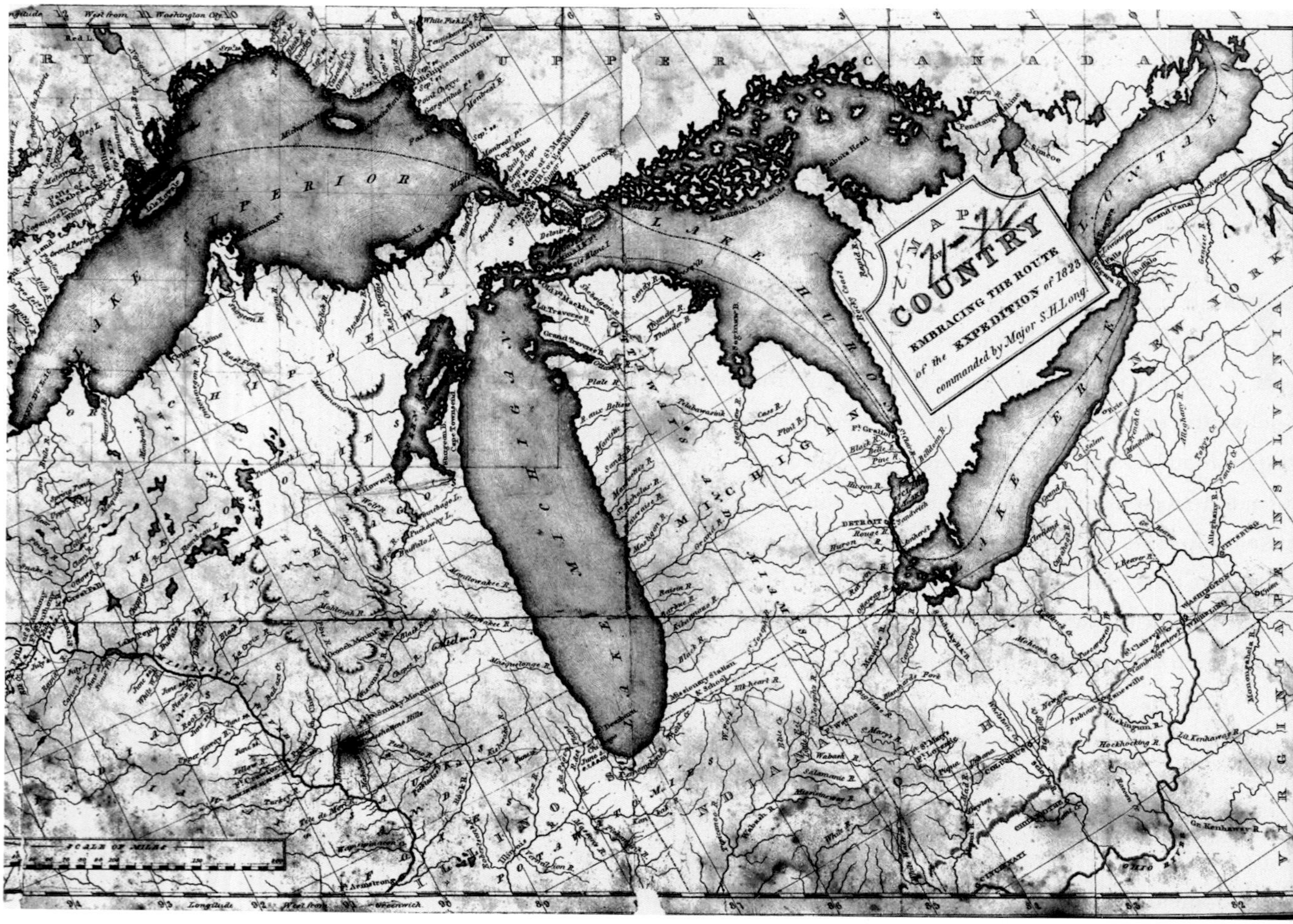


*Map depicting the route of Stephen Long's 1823 survey of the upper Midwest*

*National Archives*

as Geographers during the American Revolution and as Topographical Engineers during the War of 1812 and thereafter, were added to the peacetime Army. Unlike the other officers of the Corps of Engineers, whose primarily military duties centered on the construction and maintenance of fortifications, "topogs" performed essentially civil tasks as surveyors, explorers, and cartographers. In 1818, the War Department established the Topographical Bureau under Major Isaac Roberdeau to collect and store the maps and reports of topographical operations. Like the topogs, who numbered only six at this early date, the bureau was placed under the Engineer Department.

Almost from the outset, there was a great demand for the skills of the Topographical Engineers. The accelerated movement of Americans into the interior of the continent served to emphasize the Nation's need for networks of transportation and communication. Congress recognized the compelling nature of the requirement in 1824 and passed the General Survey Act. This law, which authorized surveys for a national network of internal improvements, became the basis for topog involvement in the development of canals, roads, and later, railroads.

Along with the growing importance of the topogs came increases in their numbers and improvements in the organizational structure. Most of the changes came during the first decade of Colonel John J. Abert's tenure as Chief of the Topographical

*John J. Abert*
*Historical Society of Pennsylvania*

Bureau. A strong-willed and ambitious West Pointer who received the appointment after Roberdeau died in 1829, Abert sought independence

*Joseph Nicollet's map of the confluence of the Minnesota and Mississippi rivers*
*National Archives*

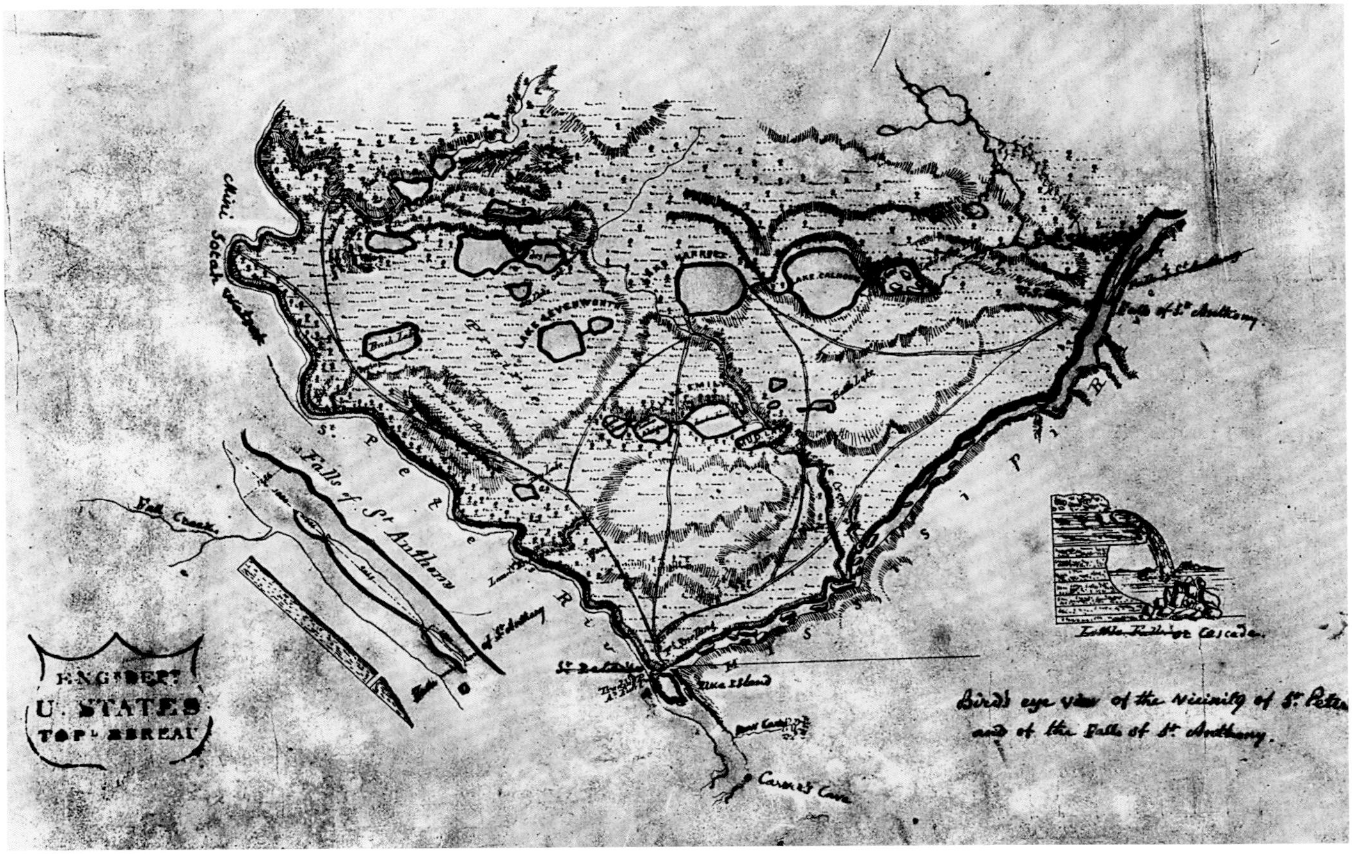

for both the bureau and the topogs. He realized the first goal in 1831 when Congress removed the bureau from the Engineer Department and gave it departmental status under the Secretary of War. Seven years later, he attained the second objective and became Chief of an independent Corps of Topographical Engineers, a position he held for twenty-three years.

Colonel Abert sought a great deal more for the topogs than prominence within the bureaucracy. While Roberdeau had been content to manage the office as a depot for maps and instruments and as a clearinghouse for correspondence, Abert saw his role as a planner and administrator for national policy regarding internal improvements and western exploration. As a member of the Board of Engineers for Internal Improvements, established to evaluate projects considered under the General Survey Act, Abert had a part in the selection of tasks and their execution. In western exploration, though, which for many years took a back seat to internal improvements, Abert's role remained minor. His bureau distributed instruments, collected maps, and forwarded correspondence.

Individual members of the Corps of Topographical Engineers, however, achieved great importance in western exploration and surveys. During the expansionist era of the 1840s, from the first stirrings of Oregon fever in the early years of the decade to the acquisition of the huge southwestern domain after the Mexican War, topogs examined the new country and reported their findings to a populace eager for information about the lands, native peoples, and resources of the West. Best known of all was John C. Frémont, the dark-eyed and flamboyant pathfinder who led three parties to the Rockies and beyond during this age of expansion. The ranks also included William H. Emory, author of a perceptive assess-

*(pictured above) Rare Corps of Topographical Engineers Model 1839 pattern uniform coat, from the USACE museum collection. This example is believed to be the uniform of First Lt. Jacob E. Blake.*

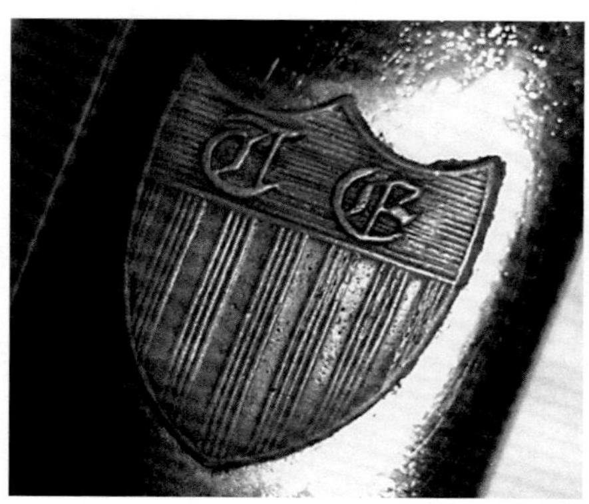

*(pictured right) Sword hilt insignia of the Corps of Topographical Engineers*

*(pictured far right) Corps of Topographical Engineers cap device*

ment of the Southwest, and James H. Simpson, discoverer of the ruins of the ancient Pueblo civilization of New Mexico. Howard Stansbury, whose report of an exploration of the Great Salt Lake is still considered a frontier classic, also wore the gold braid of the Corps of Topographical Engineers. In the 1850s, when the emphasis shifted from reconnaissance to more detailed exploration and roadbuilding, topogs continued to make their marks. John N. Macomb laid out the basic road network of New Mexico, George H. Derby initiated harbor improvements in California, and Joseph C. Ives became the first Anglo-American to descend the Grand Canyon.

The disparity between the renown of members of Abert's Corps and the obscurity of his bureau was due to the absence of a government policy regarding exploration. The Topographical Engineers frequently went into the new country on an *ad hoc* basis at the behest of a politically powerful figure like Missouri Senator Thomas Hart Benton or to accompany a military expedition. From Major Stephen H. Long's 1819 journey up the Missouri River as a minor adjunct of Colonel Henry Atkinson's Yellowstone Expedition

Lithograph, c. 1850s, depicting John C. Frémont ascending Snow Peak

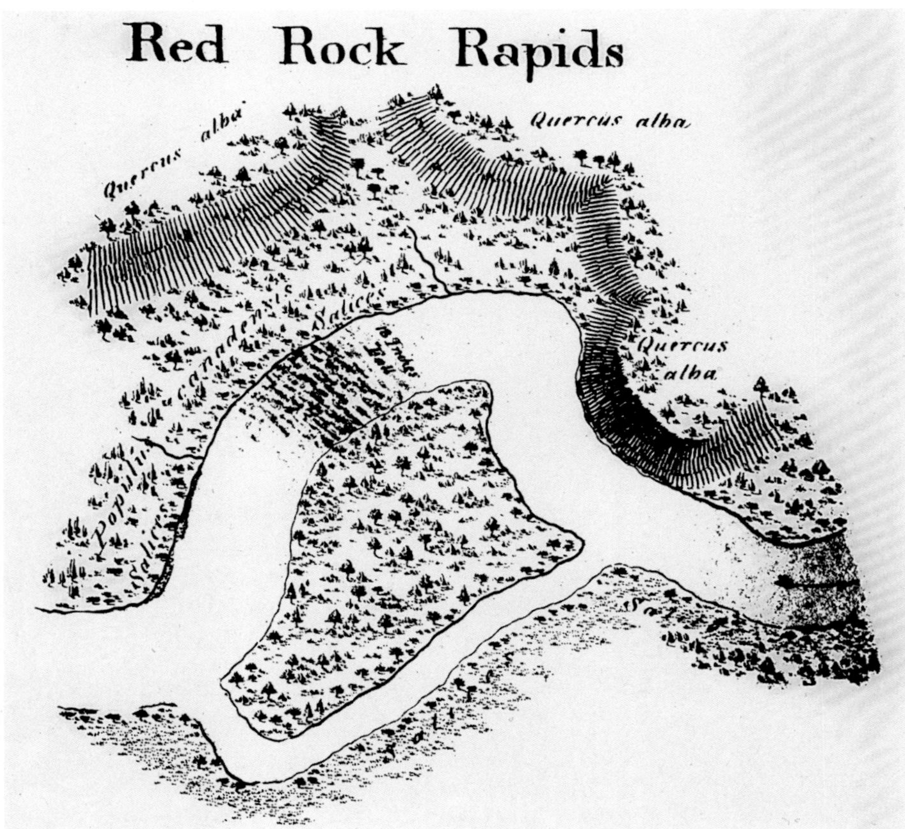


A portion of John C. Frémont's 1841 map of the Des Moines River
National Archives

to Emory's southwestern exploration during the Mexican War and Mexican boundary surveys, topog exploration often took a secondary position to other purposes.

When exploration and surveys in the Trans-Mississippi West were finally organized and coordinated in the 1850s, Abert no longer wielded the political influence that had brought his ambitions so near fruition in the 1830s. Duties he hoped would devolve on the Corps of Topographical Engineers went instead to the Office of Pacific Railroad Explorations and Surveys, a small organization created by Abert's political foe, Secretary of War Jefferson Davis. This new office would manage the surveys for railroad routes to the Pacific Ocean.

Despite the lack of a unified policy and central direction, the history of topog expeditions forms a coherent entity. Topographical officers provided the necessary link between the first explorations of the mountainmen—those rude, brawling beaver trappers, who first probed far beyond the frontier and were no less than walking storehouses of geographical knowledge—and the civilian scientific specialists, who undertook a rigorous study of western natural history and resources after the Civil

*Pacific Railroad survey party camped in the Mohave Valley*

War. Between the trappers and the specialists of the United States Geological Survey, topogs provided the Nation with an overall picture of the Trans-Mississippi region. They explored bits and pieces, as opportunity allowed, until a coherent general understanding of western topography emerged in the form of Lieutenant Gouverneur K. Warren's map of 1858. His achievement, the first accurate, overall depiction of the Trans-Mississippi West, was a milestone in American cartography. Thereafter, topog activity centered on filling in the few blank spaces in Warren's map.

During the Civil War, the Corps of Topographical Engineers was merged into the Corps of Engineers, whose officers renewed the topogs' efforts after Appomattox. Their work continued until 1879, when primary mapmaking responsibilities passed from the Army to the newly established U.S. Geological Survey. By then, the officer-explorers had done their major task. They had extended and codified the knowledge of the mountainmen and, in turn, laid the groundwork for scholarly analysis. The Topographical Engineers had performed an essential service to a nation growing in size and in self-understanding.

*Gouverneur K. Warren*

*National Archives*

# Toward a National Transportation Network: The Pacific Railroad Surveys

*Henry L. Abbot as a general officer*

By 1853, influential members of Congress had decided to support the construction of a transcontinental railway; however, there was a serious dispute over the proposed route for such a line. Congress amended Army appropriations to fund the reconnaissance of several potential routes by the Corps of Topographical Engineers. The Secretary of War, Jefferson Davis, established the Office of Pacific Railroad Explorations and Surveys and appointed Captain Andrew A. Humphreys of the Topographical Engineers to oversee the project. Ultimately, the topogs explored four different routes in seven different expeditions. The northernmost expedition, led by Isaac Stevens, a former engineer officer, traversed from Minnesota to Washington. Captain John Gunnison surveyed the area along the Arkansas Valley into the Great Salt Lake. Lieutenant Amiel Whipple explored the area along the 35th parallel through New Mexico. Two expeditions, those under Lieutenants John Pope and John G. Parke, surveyed the final route through the recent Gadsden Purchase and Texas. Additional survey parties under Robert S. Williamson and Lieutenant Henry L. Abbot and another by Parke probed the mountains of Oregon and California for railroad passes.

These parties faced an assignment of considerable complexity. Each expedition was required to report on the numerous determinants of railroad construction, among them were distances, grades, mountain passes, canyons, bridge sites, and tunnels. In addition, each survey had to consider natural resources, particularly timber, stone, coal, and water, all crucial for building and operating a railroad.

The surveying parties faced great hardships as they made their way westward. In the Northwest, the Stevens expedition ran into the blizzards of the Rocky Mountains. Pope and his men would spend many days without water on the barren Llano Estacado. The party of Abbot and Williamson stumbled into nests of rattlesnakes near Lake Klamath. In eastern Utah, Gunnison and several assistants were cut down in a predawn attack.

In spite of the obstacles, the topographic expeditions brought back a remarkable amount of data. The thirteen-volume final report was a comprehensive record of the trans-Mississippi region's flora and fauna, geological morphology, and geographical characteristics. The immense compendium of this report remains as a reference that naturalists continue to consult.

Although Congress, divided by sectional animosities, failed to agree on any one route, the surveys ultimately proved of great significance. When the first transcontinental railroad, the Union Pacific–Central Pacific running from Omaha to Sacramento, was built after the Civil War, it followed the path surveyed by Gunnison's party after his death. Later lines also went along routes first examined by these Topographical Engineers. The Pacific railroads bound together the farms, markets, resources, and industry of a growing nation.

*The Williamson survey party at work near Livermore Pass*

# 2 Contributing to National Development

*Construction of rock and brush wing dams on the Mississippi River in 1891. The photographer, Henry Peter Bosse (1844–1903), worked as a civilian engineer and draughtsman for the Corps of Engineers during the reshaping of the Mississippi River for modern transportation. Rediscovered only in the early 1990s, Bosse's photographs have won international acclaim, earning him a place among the J. Paul Getty Museum's "38 Photographers of Genius."*

# The National Road

As pioneers and immigrants settled west of the Appalachian Mountains, Americans felt a pressing need for reliable transportation routes to the newly formed states in the Ohio and Mississippi river basins. President Jefferson's Secretary of the Treasury, Albert Gallatin, and others proposed many road and river improvement projects to meet this need, but before 1840, only one project received substantial federal financial support. This was the National Road between Cumberland, Maryland, and Vandalia, Illinois, which the government built between 1811 and 1841 at a cost of more than $6 million.

Gallatin's Treasury Department supervised the construction of the first segment of the road, built between 1811 and 1818 between Cumberland on the Potomac River and Wheeling on the Ohio River. The U.S. Army Corps of Engineers assumed supervision of the road's construction in 1825, when Congress authorized the continuation of the road west of the Ohio River. The Secretary of War then ordered that the road be constructed using the method introduced in England by John McAdam. McAdam found that applying three successive three-inch layers of broken stone above ground level produced a well-compacted road surface that could bear the heaviest contemporary loads. Civilian superintendents reporting to the Engineer Department oversaw the road's construction until Congress, in 1832–1834, mandated that engineer officers be placed in immediate charge.

By then, the road east of the Ohio River had fallen into serious disrepair and Congress ordered that an engineer officer fix it and then turn it over for maintenance to the states through which it passed. That

*Conestoga wagons crossing the Appalachian Mountains on the National Road. Carl Rakeman, an artist with the Bureau of Public Roads, painted this image in the mid-twentieth century.*
*Federal Highway Administration*

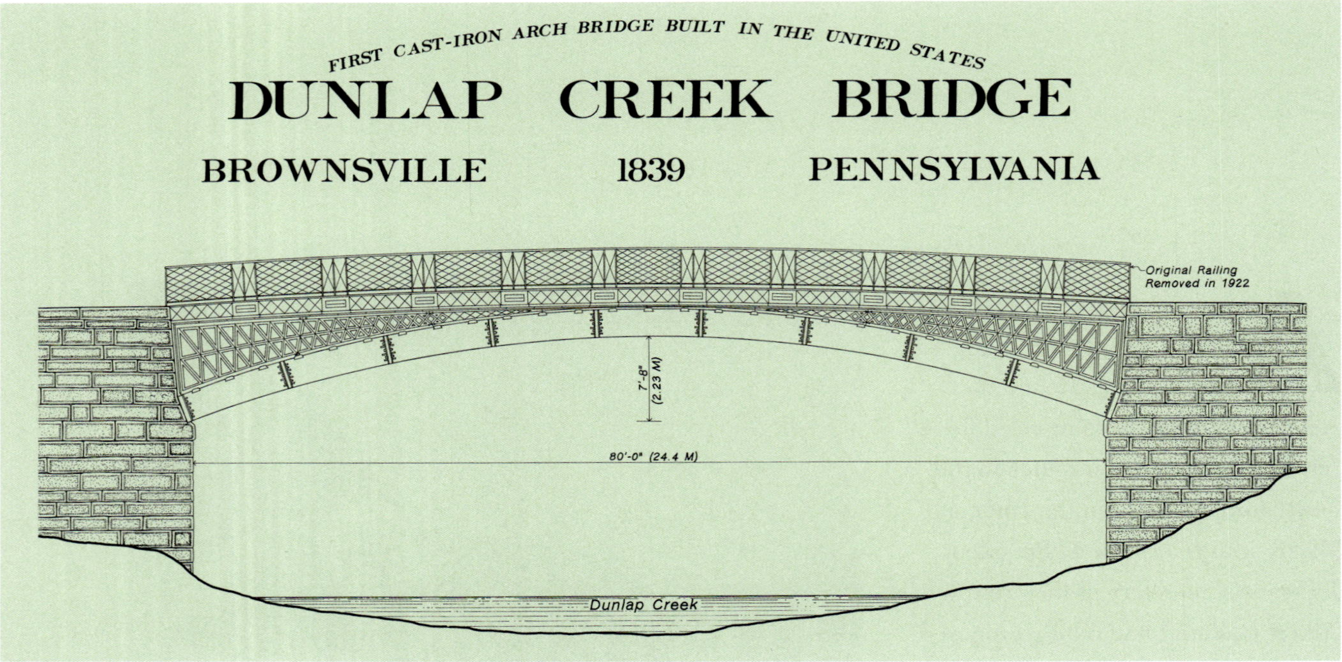


*North elevation of Dunlap Creek Bridge from drawings made in 1992 for the Historic American Engineering Record. After four other bridges failed at Dunlap Creek, Capt. Richard Delafield designed this cast-iron bridge in the mid-1830s. Construction began in 1836, but bad weather, labor shortages, and inadequate funding delayed completion until 1839.*

*Library of Congress*

section of the road had been built with large foundation stones, and many of these had worked their way to the surface at dangerous angles. In return for subsequent state assumption of maintenance responsibilities, the federal government agreed to macadamize the road, to build a new route just west of Cumberland that avoided a steep mountain ridge, and to replace several decaying original bridges.

Engineer Captain Richard Delafield, a future Chief of Engineers, supervised most of the eastern repair work. His new solid masonry bridge over Will's Creek west of Cumberland had two elliptical arches, each spanning 59 feet and standing more than 26 feet above the water. With wing walls, its total length was 291 feet. Across Dunlap's Creek at Brownsville, Pennsylvania, Captain Delafield built the first bridge with a cast-iron superstructure in the United States, an 80-foot-long span that remains in use today. The Cumberland Road Project was an early example of the Corps providing imaginative and durable engineering work under challenging circumstances.

*First cast-iron arch bridge built in the United States. Carl Rakeman painted this image of the Dunlap Creek Bridge.*

*Federal Highway Administration*

*The National Road at Clarysville, Md.*

*National Archives*

# The Corps Helped Construct Portions of Afghanistan's National Road

During the 1960s, the U.S. Army Corps of Engineers oversaw a program to improve Afghanistan's poor system of roads. At the time, Afghanistan's rudimentary highway system consisted of a 1,700-mile circle of rock-bed and dirt roads linking principal towns and cities. From Kandahar in the south, the roads ran both northeast to Kabul and northwest to Herat. The main road then looped across the northern tier of the country to connect Herat and Kabul. Spurs from this great elliptical route known as the ring road extended toward Iran to the west and Pakistan to the southeast. The Mediterranean Division's Gulf District established an Afghanistan Area Office at Kabul to tackle the construction challenge.

In 1961, the Corps initiated construction of one part of the highway system, a ninety-six-mile spur from Kandahar southeast to the border with Pakistan at Spin Baldak. Although this project had been completed in a relatively rapid manner, the major portion of the Afghanistan highway, the 300-mile road from Kabul to Kandahar, languished in the design stage. A border closing restricted construction operations for several years as the contractor had to develop alternate routes for transporting equipment and supplies—principally through Iran to Meshed, across a primitive road to Herat, Afghanistan, and on to Kandahar. But modified specifications allowed the contractor to complete the initial segment in the north by 1964. Within two more years, the contractor turned over the final portion of the highway, along with a series of bridges and drainage ditches completed as ancillary projects.

The Mediterranean Division oversaw construction of a third highway segment running seventy-five miles west from Herat to the city of Islam Qala on the Iranian frontier, which was completed by late 1967. The total cost of constructing this road had risen to $9.5 million with the considerable repair and redesign necessitated by massive flooding.

The Herat-Islam Qala highway, the Kabul-Kandahar highway, and the Kandahar-Spin Baldak highway linked systematically with Russian-built roads. The total American contribution to this highway system consisted of more than $55 million for construction and another $25 million in related costs.

The Corps played a major role in providing Afghanistan with a modern highway system at the height of the Cold War. After the Soviet invasion of Afghanistan in 1979, the road network continued to degrade while it was used to facilitate occupation of the country. With the establishment of a new national government following the overthrow of the Taliban regime in 2001, the Corps resumed its role, constructing bridges and providing technical assistance to the Agency for International Development's transportation reconstruction program in Afghanistan.

*Laying highway asphalt along the ring road*

# Lighthouses

As early as 1716, private parties built lighthouses on the Atlantic Coast. U.S. Army engineers began supervising lighthouse construction in 1827. In 1831, the Treasury Department placed funds appropriated for lighthouses in the hands of the Chief Engineer of the U.S. Army Corps of Engineers. A federal Lighthouse Board, created in 1852, assumed responsibility for supervising lighthouse construction and inspection. Three engineer officers were members of the original Lighthouse Board, and U.S. Army engineers were assigned to each of the twelve lighthouse districts.

In the nineteenth century, engineer officers designed lighthouses to help mariners weather violent Atlantic storms. Adopting European technology, those officers often innovated to solve particular problems. Major Hartman Bache borrowed from British engineers the design for the first screw-pile lighthouse in the United States. This type of pile was ideal for the bottom of the Delaware Bay because it could be securely twisted into an unstable sea floor. To fend off the floating ice that threatened a structure at Brandywine Shoal, Delaware, Major Bache installed a fence of screw piles five inches in diameter around the lighthouse. He then added an outer fence and erected a platform over the space between the two fences. Tons of stone riprap were dumped around the structure to provide additional protection.

*Cape Lookout Lighthouse, N.C., completed by the Corps of Engineers in 1859*

*Minot's Ledge Lighthouse on the Mass. coast under construction in August 1859*

*National Archives*

*Sombrero Key Lighthouse designed and built by Lt. George G. Meade. Photograph taken 1971.*
*U.S. Coast Guard*

Engineering advances later made it possible to erect sturdy lighthouses on the reefs around the Florida Keys. The most famous of these was the Sombrero Key Lighthouse, built by Lieutenant George G. Meade seven years before he met General Robert E. Lee at Gettysburg in July 1863. U.S. Army engineers also erected the first lighthouses on the Pacific Coast. By the Civil War, engineer officers

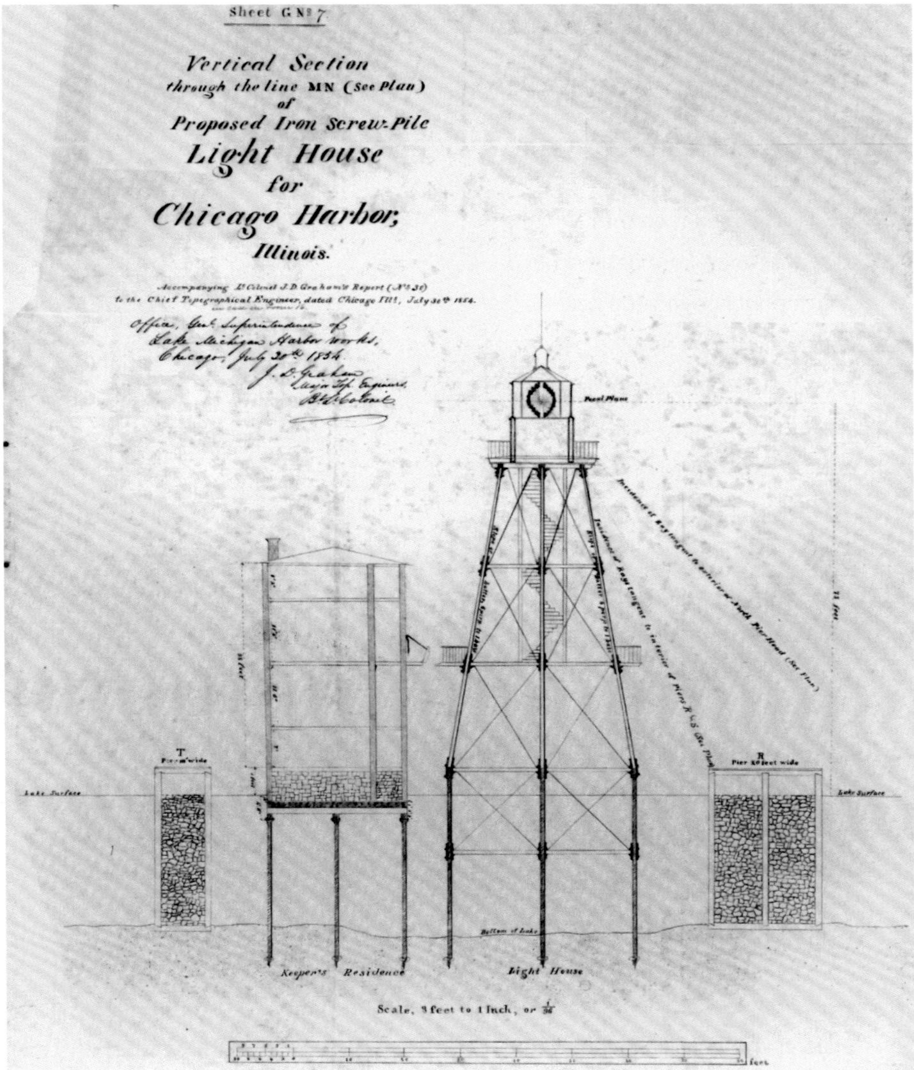


*Lighthouse plan for Chicago Harbor*
*National Archives*

had placed new Fresnel lenses in all lighthouses.

In addition to making design innovations, the Lighthouse Board oversaw significant advances in optics, sounding mechanisms, and mariner warnings. Engineers continued to serve as board members and as lighthouse district inspectors and engineers until Congress abolished the Lighthouse Board in 1910. The overall number of aids to navigation, including lighthouses, buoys, and fog signals, had grown from around four hundred at the inception of the board to just fewer than twelve thousand at its conclusion. After the board was abolished, U.S. Army engineer officers continued to work on intermittent lighthouse assignments.

# Two Lighthouses at Minot's Ledge

Captain William H. Swift designed the screw pile iron lighthouse for Minot's Ledge outside Boston harbor and construction began in 1847. He studied examples of these new technologies in England and adapted them for American lighthouses. The Minot's Ledge location was a small, rocky island battered by the sea. Swift designed the lighthouse seventy feet high with a twenty-five-to-thirty-foot base. In 1849, after a violent storm, he began to add diagonal bracing to strengthen the structure, but, before this adaptation was complete, a tremendous gale in April 1851 destroyed the structure killing two lighthouse keepers. Accusations and recriminations began immediately with critics favoring a traditional heavy stone structure. Swift asserted that modifications made by the lighthouse keeper had weakened the iron structure. Congress eventually funded a stone lighthouse that exists today, but engineers continued to build iron lighthouses safely in other locations.

The failure of Minot's Lighthouse did not stop work on the other [iron] lighthouses authorized by Congress in 1847 and had relatively little impact on their designs, apart from underscoring the importance of large bases relative to height and the need for diagonal bracing. An interesting feature of the early iron skeleton lighthouses is how unique each one was. Differences due to varying site features and requirements for the light—such as height, type of foundation, and environmental conditions (breaking waves or harbor rollers)—are understandable. But the towers varied in other ways, which indicated that the designers were experimenting. It also shows that Colonel [John J.] Abert [Chief of the Corps of Topographical Engineers] gave his officers latitude to experiment and to use their judgment in deciding the details of the lighthouses in their districts.

Quoted from Sara E. Wermiel, *Army Engineers' Contributions to the Development of Iron Construction in the Nineteenth Century,* Essays in Public Works History, #21 (Kansas City, Mo.: Public Works Historical Society, 2002), p. 18.

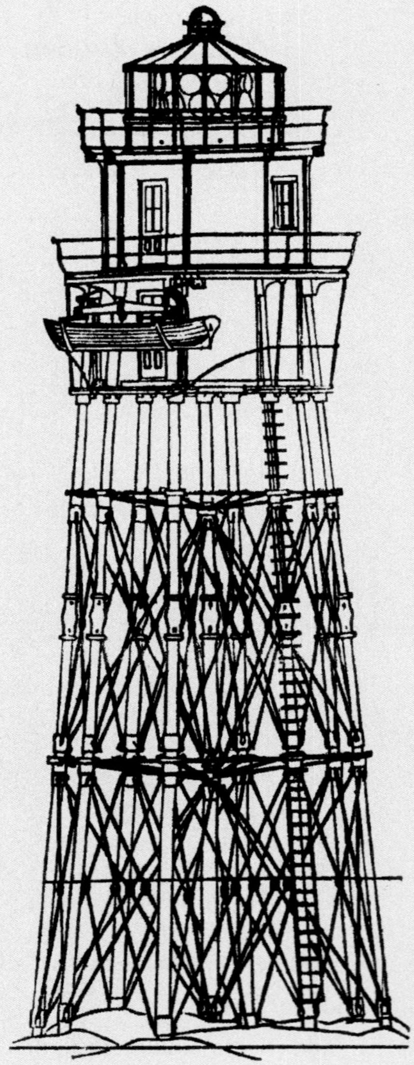

*The iron lighthouse at Minot's Ledge, Cohasset, Mass.*

Boston Journal, *April 21, 1851*

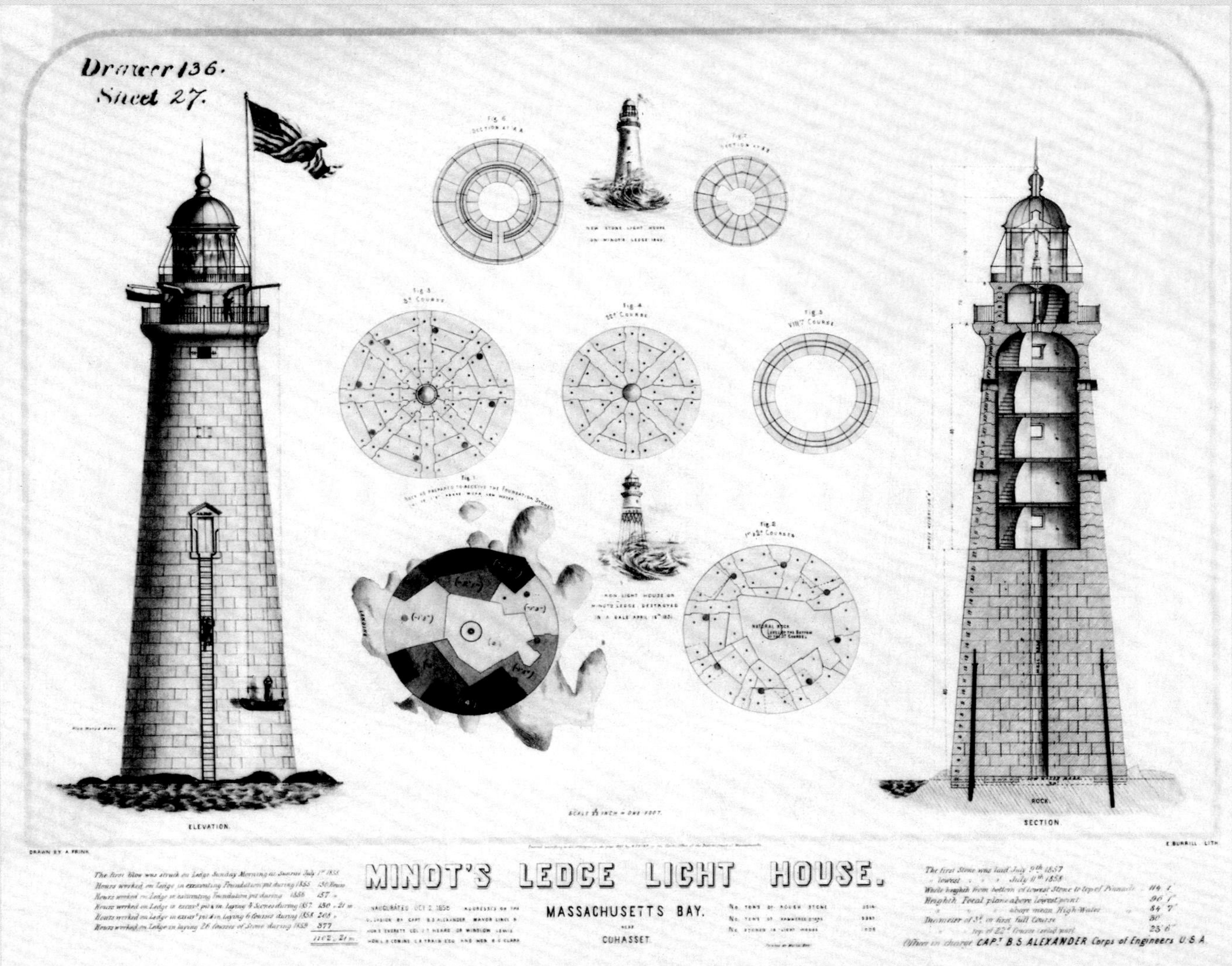


Plans for stone lighthouse at Minot's Ledge built after the destruction of the iron lighthouse in 1851

National Archives

U.S. Snagboat No. 2, *from* Harper's Weekly, *November 2, 1889*

# Origins of Civil Works Missions

One of the major lessons of the War of 1812 was that the Nation needed an improved defense and transportation system. The British had invaded the United States from the north, from the south at New Orleans, and from the east, marching inland and even putting the capital to the torch. In the 1816 mobilization studies based on the lessons of the War of 1812, the U.S. Army Corps of Engineers reported that national defense should rest upon four pillars: a strong Navy at sea; a highly mobile regular Army supported by reserves and National Guard; invincible defenses on the seacoasts; and improved rivers, harbors, and transportation systems that would permit rapid armed concentration against an invading enemy, and swifter, more economical logistical lines.

In 1819, John C. Calhoun, then Secretary of War, recommended that the U.S. Army Corps of Engineers be directed to improve waterways navigation and other transportation systems because such civil works projects would facilitate the movement of the U.S. Army and its materials while contributing to national economic development. "It is in a state of war when a nation is compelled to put all of its resources ... into requisition," said Calhoun, "that its Government realizes in its security the beneficial effects from a people made prosperous by a wise direction of its resources in peacetime."

Congress finally accepted Calhoun's recommendations in 1824. It passed the General Survey Act on April 30, authorizing the president to use U.S. Army engineers to survey road and canal routes "of national importance, in a commercial or military point of view." A few weeks later, on May 24, Congress appropriated $75,000 for improving navigation on the Ohio and Mississippi rivers. This law allowed the president to employ "any of the engineers in the public service which he may deem proper" for the work. Also under this act, the Corps began to remove snags and floating trees from the Ohio and Mississippi rivers and to improve the Ohio's channel by attacking the sandbars that impeded river commerce.

*John C. Calhoun, Secretary of War, 1817–1825, by John Wesley Jarvis*
*U.S. Army Collection*

Captain Henry M. Shreve Clearing the Raft from Red River, 1833–38, *painted in 1969 by Lloyd Hawthorne*

*Courtesy of the R. W. Norton Art Gallery, Shreveport, Louisiana*

By 1829, U.S. Army engineers were using snagboats developed by the famous steamboat captain, Henry M. Shreve, to remove obstructions in river channels. Appointed by the secretary of war as superintendent of western rivers, Shreve realized that the use of a steam engine and other design techniques would cut the cost of snag removal in half. His first double-hulled snagboat, the *Heliopolis,* successfully removed extensive obstructions along the lower Mississippi and Red rivers (and later the Missouri, Ohio, and upper Mississippi rivers). An iron

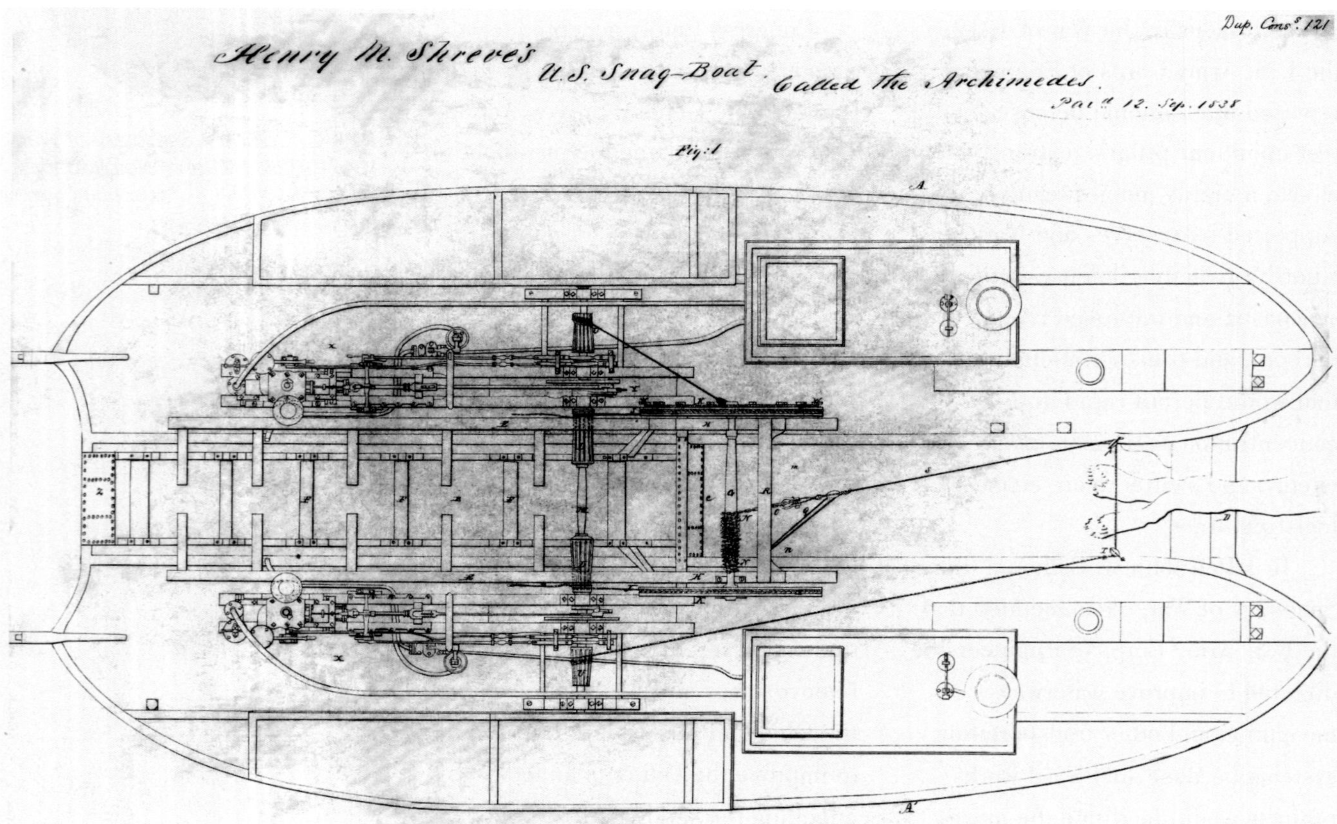


*Design plans for Shreve's snagboat* Archimedes *(1838)*

beam connecting the two hulls was used as a battering ram to dislodge a snag from the river bed. The vessel's lifting capability was provided by machinery instead of by hand, which made it a much more powerful snag remover. These Corps snagboats, which could lift a submerged tree weighing seventy-five tons lodged up to twenty feet deep, became known as "Uncle Sam's Toothpullers." Shreve, who eventually received a patent on his snagboat design, also began clearing riverbanks to prevent falling trees from becoming navigational hazards.

This early activity marked the beginning of the Corps' civil works mission—a dual role that emphasized a practical blending of civil works and military skills and fostered the development of a federal agency prepared to shoulder the engineering burden in the event of war or national emergency.

*(right, top to bottom) Henry Bosse photo of the snagboat* General Barnard *(1885), named after brevet Major General John Gross Barnard, chief engineer of the Washington, D.C., defenses during the Civil War*

*Rock Island District, U.S. Army Corps of Engineers*

*U.S. Snagboat* Chauncey B. Reese, *built in 1879*

# Long's Steamboat

Major Stephen H. Long, an engineer officer famous for his exploration of the American West and for the survey and construction of early American railroads, also designed his own steamboat. In 1818, Long planned the building of the experimental craft, the *Western Engineer,* to transport himself and a task force of scientists, naturalists, and artists as far west as possible by water on their projected trip into the frontier.

The result was a steamboat designed to navigate narrow, shallow, snag-littered channels of inland rivers. It contained a particularly strong engine to provide increased power for pushing against swift currents. Another novel feature was a paddlewheel built into the stern to reduce the danger of damage from snags. The shallow-draft boat had a seventy-five-by-thirteen-foot hull with the weight of the machine carefully distributed to permit increased maneuverability in shallow channels.

The *Western Engineer* made an imposing debut when launched on the Ohio River in May 1819. To protect the vessel from Indian attack, Major Long installed a bulletproof pilot house. In addition, he had a cannon mounted on the bow, placed howitzers along the sides, and armed the crew with muskets and sabers. The boat had a serpent-like shape to frighten any would-be attackers. Drawing but nineteen inches of water compared to the five or six feet of most steamboats, the *Western Engineer* became the prototype of the Western river steam vessels.

At the beginning of that summer, the *Western Engineer* joined the "Yellowstone Expedition" of Colonel Henry W. Atkinson. In this vessel, Long and his crew explored the Ohio River, ascended the Mississippi River, then entered the Missouri River well into Nebraska. At this point, Long abandoned the *Western Engineer* and struck out overland for the Rocky Mountains in the spring of 1820, finally reaching the Arkansas River late that summer. Though plagued throughout the expedition by frequent breakdowns, Long's steamboat was the first such vessel to explore the territory of the Louisiana Purchase and had traveled further westward than any other steamboat.

Engineer Cantonment and Western Engineer *(1820) by Titian R. Peale*

*American Philosophical Society Library*

# Waterway Development

Benjamin Henry Latrobe, a famous early nineteenth century engineer, once remarked that "nothing is so easily converted to a civil use as the science common both to the profession of a civil and military engineer." Few of Latrobe's contemporaries questioned this observation; engineers were also scientists, and navigation improvements required a scientific approach using principles developed mainly in Europe. At West Point, U.S. Army engineers learned the principles and applied them in their surveys of navigable rivers, often making their own significant contributions to river hydraulics in the process. In the early 1820s, U.S. Army Corps of Engineers officers surveyed both the Ohio and Lower Mississippi rivers. In the succeeding years, the Corps investigated a number of additional rivers. Many early navigation improvements resulted from trial and error, rather than from strict adherence to theory. If the obvious did not work, the less obvious was used until some method produced the desired result. A good example was the work on the Ohio River.

In 1824, Chief Engineer Alexander Macomb dispatched Major Stephen H. Long to the Ohio to initiate experiments on providing safer navigation. The main challenge was to deepen channels across sand and gravel bars. Major Long decided to perform experiments on a compacted gravel bar near Henderson, Kentucky, just below the mouth of the Green River. At low-river stage, this bar was covered by only fifteen inches of water. After preliminary studies, the major outfitted several flatboats with hand-powered pile drivers and

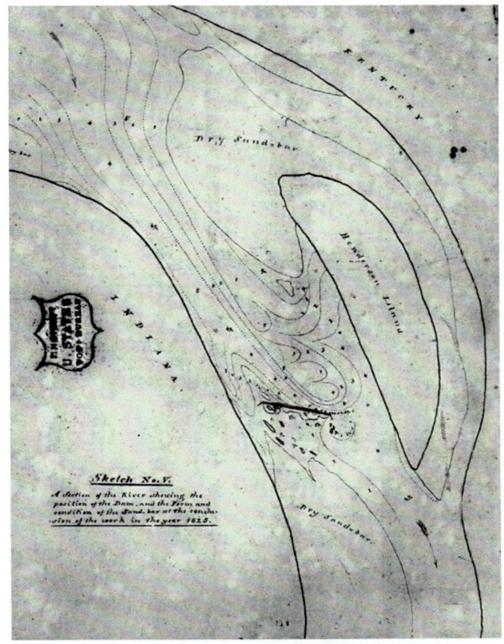


*Topographical map of Henderson Island sandbar in the Ohio River, 1825*

*National Archives*

*Early steamers on the Ohio River, c. 1820*

Up the Heights of Fame and Fortune, *Frederick B. Read, 1873* began to build a wing dam, so called because the structure extended from the bank of the river at a forty-five-degree angle. The dams decreased the width of the channel, thereby increasing the current's velocity and directing its force against the riverbed. Theoretically, this would cause the river to scour a deeper channel. Major Long built the dam to various widths, lengths, and heights. The final structure was 402 yards long and consisted of twin rows of 1,400 piles joined with stringers and filled with brush. Sediment gathered against the dam and helped anchor it to the riverbed. The project's total cost was less than $3,400.

*Henry Bosse photo of wing dams below Nininger, Minn., on the Upper Mississippi River in 1891. These river structures were designed to constrict the river at shallow places, resulting in a narrowing and deepening of the channel.*

Wing dams such as Long's were used on the Ohio and other major rivers during most of the nineteenth century, but their effectiveness was always marginal. They were easily destroyed and did not always produce the desired results. After the Civil War, Corps officers grew increasingly skeptical about the dams. Brevet Major General Gouverneur K. Warren, a well-respected engineer officer, candidly wrote in 1867, "I do not believe the country will ever stand such a heavy continuous outlay as the wing-dam system of the Ohio has caused, and I believe that the extravagant and useless expenditure there, in the palmy days of western river improvements between 1830 and 1844, did more than anything else to bring the whole subject into disrepute."

Major General Warren's pessimism was unjustified, for both Congress and commercial interests continued to support waterway improvements after the Civil War. Indeed, the support increased. River and harbor work jumped from about $3.5 million for 49 projects and 26 surveys in 1866 to nearly $19 million for 371 projects and 135 surveys in 1882. Nevertheless, Warren's frustration was shared by other engineers. W. Milnor Roberts, a well-known civil engineer, concluded in 1870 that existing navigation facilities on the Ohio, while certainly of public benefit, were no better than an "amelioration of the present difficulty." He proposed instead to canalize the river through the construction of 66 locks and dams. This project would offer six-foot slackwater navigation from Pittsburgh, Pennsylvania, to Cairo, Illinois.

Chief of Engineers Andrew A. Humphreys organized an Army Engineer Board of Inquiry, composed of Majors William E. Merrill and Godfrey Weitzel, to examine the question of canalizing the Ohio. The officers agreed with Roberts that a system of locks and dams would best provide for future navigation. Somewhat surprisingly, the recommendation met resistance from the very group that would most profit from its implementation. Coal shippers, in Major Merrill's words, were "absolutely opposed to a slack-water system, unless arrangements can be made to pass their fleets through without stopping and separating for the passage of locks."

The resistance forced Merrill, who was in charge of Ohio River improvements, to look for alternative solutions. He thought the wicket dam design, developed by Jacques Chanoine in France in 1852, might be adapted for use on the Ohio. The structure used a number of large folding boards called wickets, which were hinged to a concrete base at

*Gouverneur K. Warren as a Bvt. Maj. Gen.*

*The Davis Island Lock dedication, October 7, 1885*

the bottom of the river. Each wicket was about 3.75 feet wide and 12 feet long. When the wickets were raised, the water behind them rose high enough to permit navigation. During high water, they could be lowered to allow boats to pass unimpeded. In this way, the delays the coal shippers feared would be avoided.

In 1874, Merrill proposed that a series of movable dams, employing Chanoine wickets, be constructed on the Ohio. For the first step he recommended that a 110-by-600-foot lock and movable dam be built at Davis Island, five miles below Pittsburgh. In 1877, Congress approved Merrill's plan. A year later the Corps began construction of the Davis Island Project, completing it in seven years. The 110-by-600-foot lock was the largest in the world, as was the 1,223-foot-long dam. The dam was actually composed of 305 separate Chanoine wickets and three weirs.

*Wicket dam at Lock and Dam 52 on the Ohio River, 1996*

Impressed by the early success of the Davis Island Project, in 1888 Congress authorized the extension of the Six-foot Navigation Project down the Ohio. By 1904, two locks and dams had been completed, seven were under construction, and five more were funded. At this time, before further work was done, Chief of Engineers Alexander Mackenzie decided to conduct another complete review of the project. The basic question was whether the project should be extended down the Lower Ohio River, particularly in view of generally declining commerce on inland waterways.

Pursuant to congressional authorization, Mackenzie appointed a board headed by Colonel Daniel W. Lockwood. The Lockwood Board's review of the Ohio River Project led to recommendations for a nine-foot project for the entire course of the Ohio. This conclusion rested on the finding that the probable cost per ton-mile for a six-foot project would be nearly 50 percent greater than for the nine-foot project. In the 1910 Rivers and Harbors Act, Congress authorized the construction of a nine-foot Ohio River canalization project. The Corps of Engineers completed the $125 million project in 1929.

Meanwhile, the Corps had been busy in other parts of the country developing a reliable internal waterway system. One of the key projects, going back to the mid-nineteenth

Excavating the Illinois and Mississippi Canal, 1904

National Archives

century, was the Soo Locks at Sault St. Marie, Michigan. These locks were instrumental in securing a navigable route from the copper and iron mines on the shores of Lake Superior to the industrial plants of the East. In 1852, Congress agreed to help private interests finance the cost of building a canal at St. Mary's Falls to replace a structure on the Canadian side that had been destroyed during the War of 1812. Congress granted 750,000 acres of land to the state of Michigan. Captain Augustus Canfield of the Topographical Engineers was assigned as chief engineer and superintendent of the project for the state of Michigan. Captain Canfield's design for the canal conformed to the congressional stipulation that the passage be not less than 100 feet wide and 12 feet deep, with two locks not less than 250 feet long and 60 feet wide.

Within two decades, burgeoning traffic and larger vessels made the original canal inadequate to serve commercial needs, so Congress authorized the deepening of the St. Marys River Channel and the construction of a new facility—the Weitzel Lock. Corps work began on July 11, 1870, with an appropriation of $150,000. The original canal was

Mixing plant on the Illinois and Mississippi Canal, 1900

*Steamboats on the St. Louis waterfront, 1909*

*National Archives*

widened, and the depth was increased from 12 to 16 feet. The Corps constructed a lock 515 feet long by 80 feet wide with a lift of 17 feet.

At the time of its construction, the Weitzel Lock was considered the latest in lock technology. Its culvert valves, of the butterfly type, were operated by a single stroke hydraulic engine directly connected to the valves. Hydraulic turbines generated the power that operated the lock

*Dredging a cut-off to shorten the river at Jackson Point, Miss., 1940*

*Barge and articulated concrete revetment along the Arkansas River, 1940*

gates. A movable dam was also introduced to shut off the flow of water during maintenance operations.

The U.S. Army's success in providing a passage to Lake Superior and Canada's commitment to canal building whetted the desires of shippers and industrialists for a deep-water route through the Great Lakes—a dream eventually realized in the twentieth century with the completion of the St. Lawrence Seaway.

It was the turn of the century when Congress responded to the renewed interest in water transportation by authorizing navigation projects designed to create an integrated system connecting inland areas with coastal harbors. Sandbars and rapids along the Ohio, Missouri, Arkansas, and other major rivers posed significant obstacles to the maintenance of year-round navigation channels. Eventually, with the advancement of lock-and-dam technology and more efficient dredging equipment, a nine-foot channel depth was ensured in the Mississippi and its major tributaries.

*Completion of the dewatering of the cofferdam at the Olmstead Locks and Dam in Illinois, which took forty days, August 8, 1995*

U.S. Army Corps of Engineers navigation projects continue to play an important role in support of America's economic well-being. Commercial use of the twelve thousand miles of inland and intracoastal waterways has increased: approximately one-sixth of all intercity cargo is transported by water. Waterborne commerce, recognized by experts to be the least expensive and least energy-consumptive means of transportation, is the logical choice for shippers of energy-producing commodities. Petroleum and coal together constitute more than half of all waterborne freight on the federally maintained waterways.

This expansion of commercial water transport has been facilitated by the Corps' work on major waterways, including locks and dams. The Corps dredges more than 300 million cubic yards of material annually to maintain authorized channel depths and constructs bank stabilization projects in its traditional role as the primary developer of the Nation's waterways. Also, as of 2005, engineer districts and divisions owned 257 lock chambers at 212 sites, although only 240 chambers at 195 sites received funding and were operational. The oldest operating locks are Locks 1 and 2, which were built on the Kentucky River in 1839. The Nation's newest lock, opened in July 2004, is Montgomery Point Lock located on the White River in Arkansas. An efficient system of interconnected waterways not only provides vital commercial links, it has proven to be a key factor in America's ability to mobilize in the event of war.

*John T. Myers Lock and Dam, Ky.*

*Launching the new dredge* Essayons, *1982*

# Lieutenant Eugene A. Woodruff: "A Model for All Similar Undertakings"

The "can-do" spirit of U.S. Army engineers often manifests in unexpected contributions to the public well-being. Such devotion is exhibited in an anecdote arising from work on an early Corps navigation project. In 1873, Captain Charles W. Howell, District Engineer at New Orleans, assigned his deputy, Lieutenant Eugene A. Woodruff, to the Red River of

*Captain Charles W. Howell*

Louisiana as supervisor of the project to clear the river of the "great log raft," a formidable obstruction to navigation.

In September of that year, Lieutenant Woodruff left his workboats and crew on the Red River to visit Shreveport and recruit a survey party. When he arrived in Shreveport, he found the city in the grip of a yellow fever epidemic. Fearing that he might

*The U.S. Steamer* Aid *battles Raft Number 5 on the Red River*

*U.S. Military Academy Library*

carry the disease to his workmen if he returned to camp, Woodruff elected to remain in Shreveport and tend to the sick. Volunteering his services to the Howard Association, a Louisiana disaster relief charity, he traveled from house to house in his carriage, delivering food, medicine, and good cheer to the sick and dying. He contracted the disease and died of it in Shreveport on September 30.

Captain Howell effectively captured this spirit in his eulogy to Woodruff: "He died because too brave to abandon his post even in the face of a fearful pestilence and too humane to let his fellow beings perish without giving all the aid in his power to save them. His name should be cherished, not only by his many personal friends, but by the Army, as of one who lived purely, labored faithfully, and died in the path of duty.... His conduct of the great work on which he was engaged at the time of his death will be a model for all similar undertakings and the completion of the work a monument to his memory."

Captain Howell then assigned the task of completing the work on the Red River to Assistant Engineer George Woodruff, the lieutenant's brother. On November 27, 1873, the engineers broke through the raft, finally clearing the Red River for navigation.

*Flood refugees flee to the levees at Hickman, Ky., 1912.*

# Flood Control

Congress did not authorize a comprehensive topographic and hydrographic study of a major river basin until 1850, when floods along the Mississippi River drew congressional attention to the need for a practical plan for flood control and navigation improvements at the river's mouth. The Secretary of War, Charles M. Conrad, sent Lieutenant Colonel Stephen H. Long and Captain Andrew A. Humphreys, two officers of the Corps of Topographical Engineers, to the Mississippi Basin to conduct the survey. Charles S. Ellet, Jr., one of the best-known engineers of the day, also applied to make the delta survey. Conrad suggested that Ellet work with Long and Humphreys, but Ellet preferred to work independently. Under pressure from some congressmen and after seeing President Millard Fillmore, Conrad relented, dividing the $50,000 congressional appropriation between the U.S. Army's survey and Ellet's.

Before the U.S. Army survey was complete, Captain Humphreys became quite ill and took an extended leave of absence. Lieutenant Colonel Long drafted a report based on Humphreys' notes, but he confined it simply to an exposition of what had been done without offering any specific recommendations. Therefore, Ellet's essay became the first comprehensive study of flood control on the Mississippi. Both reports were sent to Congress in January 1852. What distinguished Ellet's submission was the author's insistence on both the practicability and value of building reservoirs on the Mississippi's tributaries to reduce flooding. That recommendation prompted Colonel John J.

*Bvt. Maj. Gen. Andrew Atkinson Humphreys during the Civil War*

*Early levee construction*

*The Humphreys-Abbot Report (1861) represented the most thorough analysis of the Mississippi River ever completed and decidedly influenced the development of river engineering in America.*

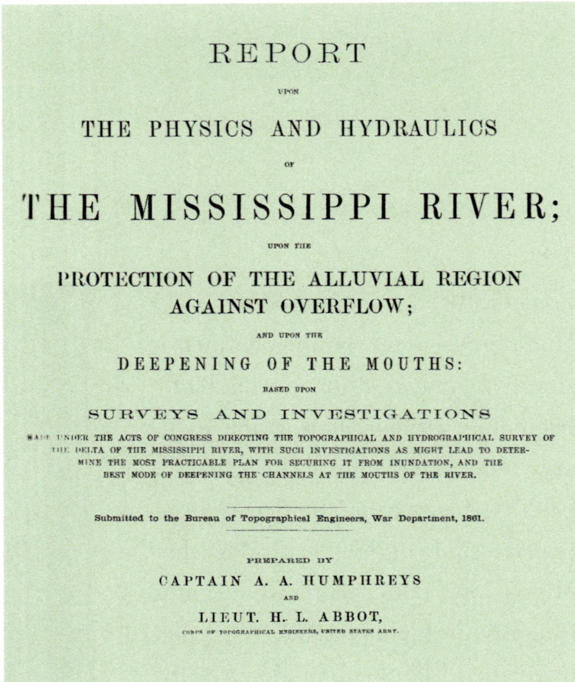

REPORT

UPON

THE PHYSICS AND HYDRAULICS

OF

THE MISSISSIPPI RIVER;

UPON THE

PROTECTION OF THE ALLUVIAL REGION AGAINST OVERFLOW;

AND UPON THE

DEEPENING OF THE MOUTHS:

BASED UPON

SURVEYS AND INVESTIGATIONS

MADE UNDER THE ACTS OF CONGRESS DIRECTING THE TOPOGRAPHICAL AND HYDROGRAPHICAL SURVEY OF THE DELTA OF THE MISSISSIPPI RIVER, WITH SUCH INVESTIGATIONS AS MIGHT LEAD TO DETERMINE THE MOST PRACTICABLE PLAN FOR SECURING IT FROM INUNDATION, AND THE BEST MODE OF DEEPENING THE CHANNELS AT THE MOUTHS OF THE RIVER.

Submitted to the Bureau of Topographical Engineers, War Department, 1861.

PREPARED BY

CAPTAIN A. A. HUMPHREYS

AND

LIEUT. H. L. ABBOT,

CORPS OF TOPOGRAPHICAL ENGINEERS, UNITED STATES ARMY.

Abert, Chief of the Corps of Topographical Engineers, to write, "While I willingly admit that all the speculations of a man of intellect are full of interest and deserving of careful thought, yet I cannot agree with him that these reservoirs would have any good or preventive effects upon the pernicious inundations of this river."

Nine years later, Humphreys elevated Abert's comment to official Corps policy. After a long convalescence and subsequent work on the Pacific Railroad Surveys, Humphreys took up his task once more in 1857, this time with the assistance of Lieutenant Henry L. Abbot. The

*Flood victims on a Mississippi River levee at Arkansas City, Ark., 1927. Note flooding behind the levee from the Arkansas River.*

Shoring up a levee near Memphis, Tenn., 1927

young lieutenant supervised a party that took gauge readings, determined discharges at various points, measured cross-sections, and reported on the state of various river improvements. When possible, he compared his data with that obtained by earlier survey parties. "In a word," Abbot later wrote, "the finger was to be firmly placed on the pulse of the great river, and every symptom of its annual paroxysm was to be noted." It was in the shadow of the Civil War that Humphreys and Abbot finally put their five hundred-page report together. They submitted it to the Chief of Topographical Engineers in August 1861, a few months after the firing on Fort Sumter. Humphreys was technically the report's author, but he insisted on listing Abbot as coauthor in recognition of his diligence and skill.

Humphreys's and Abbot's *Report Upon the Physics and Hydraulics of the Mississippi River* not only contained much new data about the Mississippi, it also analyzed other alluvial rivers around the world. The authors introduced entirely new formulations to explain river flow and sediment resistance and concluded that Ellet's calculations and assumptions were erroneous. Their own position, based on significantly more information, was that "levees only" could prevent flooding on the Mississippi. Neither reservoirs nor cut-offs were needed. Already a member of the American Philosophical Society, Humphreys received numerous honors for his work on hydraulics. He was made an honorary member of the Imperial Royal Geological Institute of Vienna in 1862 and the following year a corporator of the National Academy of Sciences and a fellow of the American Academy of Arts and Sciences. In 1864 he was elected an honorary member of the Royal Institute of Science and Arts of Lombardy, and in 1868, Harvard College conferred upon him the degree of Doctor of Laws.

In considering navigation and flood control as interrelated problems, Humphreys, Abbot, Ellet, and other engineers in the United States and many in Europe were ahead of their time. By 1879, growing pressures for navigation improvements and flood control prompted Congress to establish the Mississippi River Commission, a seven-member organization responsible for executing a comprehensive plan for flood control and navigation works on the Lower Mississippi. This permanent body of experts included three members from the U.S. Army Corps of Engineers, one from the Coast and Geodetic Survey, and three civilians, two of whom had to be civil engineers. The creation of this river basin authority marked the federal government's growing commitment to the development of a reliable inland waterway system. Initially, Congress authorized the commission to build and repair levees only if the work was part of a general navigation improvement plan. Monumental floods in 1912 and 1913, however, drew national attention to the need for federal flood relief legislation. Finally in 1917, Congress passed the first Flood Control Act. This legislation appropriated $45 million for flood control on the Lower Mississippi and $5.6 million for work on the Sacramento River.

*Flood at Greenville, Miss., 1927*

*High water at Pine Bluff, Ark., 1927*

The 1861 report of Humphreys and Abbot enormously influenced river engineering in the United States. Until 1927, when a catastrophic flood hit the Lower Mississippi, the Corps' position was that "levees only" could control flooding on the river. The Corps was not unalterably opposed to reservoirs, however. Several were built on the Upper Mississippi, but principally to aid navigation.

Advocates of reservoir construction also received support in 1897 from Captain Hiram S. Chittenden of the U.S. Army Corps of Engineers. Chittenden's essay, *Preliminary Examination of Reservoir Sites in Wyoming and Colorado*, submitted in response to a congressional directive, was a comprehensive and lucid presentation of engineering, physiographic, and economic data. In it Chittenden declared that reservoir construction in the arid regions of the West was "an indispensable condition to the highest development of that section." He also warned, "The function of reservoirs will always be primarily the promotion of industrial ends; secondarily only, a possible amelioration of flood conditions in the rivers." So far as the Mississippi was concerned, "the difficulty was not so much a physical as a financial

Capt. Hiram M. Chittenden

Floodwall at Cairo, Ill., 1936

*Corps personnel laying down sandbags*

The 1936 Flood Control Act recognized that flood control was "a proper activity of the Federal Government in cooperation with States, their political subdivisions, and localities thereof." Congress gave responsibility for federal flood control projects to the U.S. Army Corps of Engineers, while projects dealing with watershed run-off and soil erosion were assigned to the Department of Agriculture. This law made the Corps responsible for flood control throughout the Nation, working in cooperation with the Bureau of Reclamation. In the years following passage of this law, the Corps built, pursuant to congressional authorization and appropriation, close to four hundred reservoirs whose primary benefit was flood control; however, flood control alone could never have justified the construction of these reservoirs. In the age of multipurpose projects, possible navigation, water storage, irrigation, power, and recreation benefits were considered before a final economic benefit figure was determined.

Since the 1970s, in an era increasingly sensitive to environmental protection and to the limitations of traditional structural answers to flood-damage reduction, the Corps has designed and implemented hundreds of nonstructural projects to provide some level of flood protection. Nonstructural measures reduce or avoid flood damages without significantly altering the nature or extent of flooding. They may be considered separately or in combination with structural measures. Nonstructural methods include moving communities away from a flood's destructive path, raising and flood proofing buildings, acquiring vulnerable structures, preserving wetlands, buying out floodplains, and establishing a flood warning system.

*Navarro Mills Lake and Dam in Texas. Completed in 1963, the dam provided flood control and water conservation and later served an important recreational function.*

# The Bicycle Flood Fight, 1897

The Fourth Engineer District at New Orleans received word in early 1897 that a major flood was southbound on the Mississippi. Major George M. Derby, District Engineer, and civilian assistant W. J. Hardee prepared to defend the levees along more than 450 miles of river in the Fourth District. As had become customary by 1897, they stationed barges and quarterboats loaded with tools, sandbags, and lumber at roughly 15-mile intervals along the river with towboats assigned to each 60-mile section.

During previous flood emergencies, Fourth District personnel had encountered great difficulty maintaining regular patrols of the levee system and coordinating the work of five other involved parties: individual planters, railroads, parish governments, levee districts, and state government. Backwater and washouts had closed roads and railroads; there were no motorized vehicles available then, and the towboats moved too slowly and usually too far from the levees for proper inspection.

To improve coordination and inspection, Hardee equipped field personnel with bicycles. During the subsequent flood fight, the inspectors kept constantly on the move atop the levee crowns on their new transportation equipment. Hardee personally covered as much as thirty miles of levee a day on his bike, including stops for observation (and presumably to catch his breath).

*Installation of large turbine at Wilson Dam on the Tennessee River near Florence, Ala. The dam was the largest in the world upon completion in 1925.*

# Hydropower Development

Since the turn of the twentieth century, the U.S. Army Corps of Engineers has moved from a position opposing involvement in hydroelectric power to one of total endorsement. By 1900, Congress had already initiated partial federal control over dam building. The Corps participated in the regulatory process but conceived its role narrowly.

In January 1905, Brigadier General Alexander Mackenzie, the Chief of Engineers, summed up the Corps' traditional views on the federal government's limited role in improving American waterways. Congress, he said, could legally "exercise control over the navigable waters of the United States … only to the extent necessary to protect, preserve, and improve free navigation." Mackenzie further maintained that nothing should be permitted to interfere with the central purpose of locks and dams—to facilitate navigation and commerce. All other interests were clearly secondary. These views fit the prevailing judicial interpretation of federal powers under the Constitution's Commerce Clause.

During the years following Brigadier General Mackenzie's pronouncements, attitudes gradually changed. The engineers became convinced that the escalation in private dam building, largely for hydropower purposes, threatened to jeopardize their prerogatives in navigation work, and they guarded those prerogatives jealously. While the federal government redefined its part in water resources development, the Corps staked out its own territory. As an auxiliary to navigation and later to flood control, hydropower benefited from more liberal interpretations of federal authority. Cautiously, with frequent hesitation and some inconsistency, the engineers embraced the new philosophy. What began as a regulatory role in hydropower expanded to include much more. By mid-century, the U.S. Army Corps of Engineers emerged as the largest constructor and operator of federal power facilities.

The change in the engineers' role was dramatic by the end of the 1920s. By that time, they were heavily involved in surveying rivers for flood control, power, and irrigation,

*Brig. Gen. Alexander Mackenzie*

Chief Joseph Dam under construction, Wash., 1955

Contemporary view of Chief Joseph Dam along the Upper Columbia Basin

California's Folsom Dam in 1956, the year its construction was completed

as well as for navigation. Public power at multipurpose projects took hold during the New Deal and proliferated after World War II. In the mid-1950s, the Corps had more than twenty multipurpose projects under construction. By 1975, the energy produced by Corps hydroelectric facilities was 27 percent of the total hydroelectric power production in the United States and 4.4 percent of the electrical energy output from all sources. By the late 1980s, the Corps was operating and maintaining approximately seventy-five projects with hydropower facilities, and the total capacity at Corps dams was more than 20,000 megawatts.

Powerhouses, surge tanks, and switchyard at the Fort Peck Dam along the Missouri River, Mont.

The largest hydropower dams built by the Corps are on the Columbia and Snake Rivers in the Pacific Northwest. The biggest of these is the John Day on the Columbia River, which has a generating capacity of nearly 2,200 megawatts. Although by 2005 the Corps' overall percentage of hydropower capacity had slipped to 24 percent of national hydropower capacity and 3 percent of the total electrical supply, the contribution to the Nation has remained substantial.

*Fort Peck Spillway, Mont.*

*Powerhouse construction, Richard B. Russell Dam on the Savannah River, Ga., 1982*

In 1951, the Chief of Engineers referred to the development of hydropower as "one of the most important aspects of water resource development." Further, he argued, "proper provisions for hydroelectric power development are an essential part of comprehensive planning for conservation and use of our river basins for the greatest public good." Two decades later, the Office of the Chief of Engineers reaffirmed and strengthened its commitment, stating that "generation of hydroelectric power to serve the growing needs of the American people is a task the Corps welcomes." The Corps' turnabout and its expanding mission in hydroelectric power development were a significant part of the organization's history during the latter twentieth century. Today, the U.S. Army Corps of Engineers continues to operate, maintain, and occasionally add capacity at existing hydroelectric plants.

*(top to bottom)* ▶

*Generators at Bonneville Dam, Ore.*

*Bonneville Lock and Dam on the Columbia River*

# The U.S. Army Corps of Engineers: Dam Destroyers?

On January 15, 1907, Major William Sibert, Pittsburgh District Engineer, learned the depressing news that heavy flooding was undermining the abutment of Allegheny River Dam 3. If the dam continued to hold, which seemed likely, the flooding would gradually undermine the bank, thereby threatening a railroad track and a million-dollar glass factory. Already nine homes, various outbuildings, and 5.3 acres of land had caved into the river.

After a long and undoubtedly agonizing discussion with his staff, Major Sibert made his decision: the dam would have to go. To allow the water to continue around the dam was to invite further catastrophe.

The next morning blasting began. Five-hundred-pound dynamite charges were placed along the dam crest. Dynamiting continued until a 560-foot section at midstream had been removed. Then stones were placed along the bank to protect the glass factory and the railroad.

On January 30, the *New York Sun* printed an editorial that attacked the lack of progress on waterway projects; however, the editors noted, "no charge of dilatoriness can be brought against the officer who a few weeks ago saved a million dollars worth of property by assuming the responsibility of blowing up $80,000 worth of dam." Major Sibert became perhaps the only Corps officer ever commended by the Chief of Engineers for blowing up a government dam. His courage, imagination, and ability to bend to circumstances set high standards for his successors at the Pittsburgh District Office.

*Lt. Col. William L. Sibert*

*(above) View of eroded bank below abutment, Allegheny River, Dam 3, Jan. 1907*

*(below) View of broken dam, Allegheny River, Dam 3, Jan. 1907*

*(both images) Library and Archives Division, Historical Society of Western Pennsylvania, Pittsburgh, Penn.*

# The Environmental Challenge

As explorers and mapmakers for the pioneers, the engineers were among the first to recognize the need for protection of natural resources. As early as the 1840s, when the vast herds of buffalo seemed limitless to most travelers, engineer officers warned of their impending destruction. In one observation, Captain Howard Stansbury noted their shrinking ranges and warned that the buffalo "seem destined to final extirpation at the hands of men." While it is unfortunate that such admonitions very nearly came to pass, it is illustrative that at one point in time, one of the few surviving buffalo herds was protected at a U.S. Army Corps of Engineers project.

The U.S. Army Corps of Engineers was also influential in the creation and development of the first national park at Yellowstone in 1874, and the Corps operated and protected that park for many years. In the 1870s, Captain William Ludlow and an engineer survey party at Yellowstone confronted tourists, harbingers of the future, carving their initials, scattering their rubbish, and breaking off pieces of rock formations. Alarmed, Captain Ludlow pleaded with the visitors to respect nature's work. He stopped one woman, poised with a shovel over a mound formed over thousands of years by a bubbling spring's mineral deposits, in time to prevent her smashing the formation. In his report, Captain Ludlow proposed several ways to protect the new park. Congress authorized his recommendations, including military patrols and assignment of road construction to Army engineers, in 1883.

For thirty-five years, from 1883 to 1918, when the newly created National Park Service took over Yellowstone, the U.S. Army Corps of Engineers built and maintained the park's roads and bridges, including 279 miles of main roads, 25 miles of secondary roads, and 106 miles of approach roads in the forest reserves. Partly thanks to Captain Ludlow, who had provided the blueprint for saving the park, and Lieutenant Dan C. Kingman and Captain Hiram M. Chittenden, who designed and oversaw construction of a road system that has left a

*Brig. Gen. William Ludlow*

*Thomas Moran (1837–1926),* Golden Gate, Yellowstone National Park, 1893, *oil on canvas. The artist participated in the Hayden geological survey of Yellowstone in 1871 and returned to the region in 1892 to paint a view of the pass named "Golden Gate." In addition to capturing the inspired beauty of the region, Moran also depicted a precipitous section of the "Grand Loops," a system of scenic roads built under the supervision of Lt. Dan C. Kingman, an officer in the Army Corps of Engineers.*

*Buffalo Bill Historical Center, Cody, Wy.*

lasting imprint, Yellowstone became one of the crown jewels of America's scenic wonders.

To prevent the obstruction of navigable waterways, Congress in the 1870s directed the Corps to regulate the construction of specific bridges. The job was expanded during the 1880s and 1890s to prevent dumping and filling in the Nation's harbors, a program that was vigorously enforced by the engineers. At the Port of Pittsburgh in 1892, for instance, the Corps took a grand jury on a boat tour of the harbor and later obtained some fifty indictments against firms dumping debris into the harbor. When the engineers learned that firms were piling debris on the stream banks during the day and pushing it into the harbor at night, they began night patrols in fast boats with searchlights.

In 1893, a citizen of an Ohio River city complained to the Corps that the city was dumping into the river "household garbage, refuse of wholesale commission and slaughter houses, wagon loads of decaying melons, fruit and vegetables and car-

*Lake Lanier metal basket erosion control, Buford, Ga.*

casses of animals." The city officials replied that the complaint was exaggerated—very few dead animals were dumped into the river—and refused to stop the practice because the city then would have to build incinerators to dispose of the refuse. The Corps managed to stop the dumping anyway, forced the city to build an incinerator, and prosecuted the offenders, arguing that the garbage formed piles sufficient to obstruct navigation.

In the Rivers and Harbors Act of 1899, Congress gave the Corps authority to regulate almost all types of obstructions to navigation. The engineers were disappointed that they were not also given authority to deal with polluters, for many of the Corps personnel lived on the waterways and water quality was an immediate personal concern.

*Great Salt Plains Lake, Okla., 1964. Located east of the Salt Fork of the Arkansas River, the lake is the oldest in the Tulsa District.*

The Corps used the Rivers and Harbors Act of 1899 to the fullest extent legally possible to protect the environment of navigable waterways. In one extreme instance, the Corps managed to stop a firm from discharging a liquid effluent into a waterway by contending in court that the discharge obstructed navigation because it entered steamboat boilers and corroded them. The Oil Pollution Act of 1924 gave the Corps responsibility for protecting the Nation's harbors from offensive and dangerous oil discharges; however, the Corps could not adequately control the problem because of lack of regulatory power and insufficient manpower, and Corps officers periodically urged Congress to grant the agency adequate authority and resources.

The Corps' regulatory authority was expanded by the Clean Water Act (Federal Water Pollution Control Act) of 1972 to include all waters of the United States. The Corps began to regulate discharges of dredged or fill materials into any waters of the United States, and the permit program that resulted gave environmental protection the fullest consideration. This new work was well received even among strong environmentalists. One member of the National Resources Defense Council commended the Corps for the "will with which it is turning to carrying out the responsibilities Congress gave it in Section 404 for protecting the water quality on which the health and economic well-being of every American depends."

*Davis Pond Freshwater Diversion Project: lower guide levee at Lake Salvadore at Mile 150 on the Mississippi River, La., 1998*

In 1990, under Public Law 101–640, Congress officially directed the Secretary of the Army to include environmental protection as one of the Corps' primary missions. Four years earlier, in the Water Resources Development Act of 1986, Congress had authorized the Corps to review the operation of completed water resources projects to determine the need for modifications to improve environmental quality. Subsequently, in 1992 and 1996, the Corps received additional authorization to protect, restore, and create aquatic and ecologically related habitats, including wetlands. In the twenty-first century, the Corps actively promotes and is directly involved in ecosystem restoration throughout

*Kotzebue, a hub village in northwestern Alaska, was included in an environmental infrastructure assistance initiative that recommended an upgrade to its freshwater system.*

the country, focusing on water and related land resource problems.

Along with protective measures for the environment, the Corps pursues an active program for the preservation of cultural resources on its own land and at authorized project sites. The authorizing legislation, Section 106 of the National Historic Preservation Act of 1966, stipulates that up to 1 percent of the funds for a project can be expended for cultural resource surveys, for artifact and data recovery, and for mitigation efforts. The Corps' cultural resource preservation efforts have generated substantial results. For example, the Corps relocated a navigation lock on the Tennessee-Tombigbee Waterway to avoid destroying an Indian burial ground; and in Pennsylvania, the Corps successfully preserved a unique nineteenth-century wagon works by moving it from the project area. To avoid accidental destruction of archeological sites, the Corps is searching for the homes of ancient tribes, especially along proposed dredge disposal sites.

The Corps' responsibility for improving and maintaining navigation on the Nation's waterways requires dredging if channels are to remain open. In 1969, the dredging

*Restored Gruber Wagon Works, Berks County, Pa.*

*Point Mouillee confined disposal facility, Mich.*

*Gaillard Island in Mobile Bay, Ala., where a man-made dredge disposal site is home to 16,000 shore and seabirds, including the Brown Pelican.*

program was attacked as environmentally unsound. "All of a sudden, dredging became a four-letter word," reminisced Lieutenant General John Morris of the Corps. "Now this came as rather a surprise to us," he continued, "since dredging has been a daily activity within the Corps for 150 years and nobody paid much attention to it."

In 1970, the Corps began the Dredged Material Research Program to identify dredging and dredged material disposal systems that would be compatible with the new environmental protection mission. Completed in 1978, the Dredged Material Research Program reversed some traditional thinking about the effects of dredging. It indicated that dredging need not have adverse impacts on aquatic life and that dredged materials can create new wetlands and wildlife management areas. The research identified improved methods for constructing diked disposal areas and for using physical, chemical, and biological agents in the dredging process. It demonstrated that dredged fill can be used to reclaim strip-mined lands and other environmentally damaged areas.

Streambank erosion can cause major detrimental impacts on the environment and human welfare. It results in sediment deposits in reservoirs and waterways; it impairs navigation, flood control, and water supply project effectiveness; and it blights valuable recreation areas and streambank lands. Since 1969, the Corps has conducted intensive studies of streambank erosion, with demonstration control projects along the Missouri, Ohio, and Yazoo rivers, and has identified its causes and some potential new techniques for its control.

Since 1969, the Corps' Coastal Engineering Research Program has devised some innovative approaches to the problems of beach erosion, coastal storm damage, and navigation along the coastline. Analysis of wave patterns has led to rational design of rubble mound structures for the protection of threatened beaches and coastline. Research has identified possible uses for beach and marsh grasses in controlling coastal erosion and has established some basic relationships governing the size and shape of coastal inlets and harbor entrances.

Fish and wildlife conservation has been a concern of the Corps since Captain Stansbury warned that the buffalo were disappearing. The engineers built the first federal fish hatchery in 1879–1880 and have included such features as fish ladders in project planning for many years. Corps projects are designed to minimize damage to fish and wildlife resources, and even enhance wildlife resources through effective wildlife management. Approximately 2.5 million acres of land are primarily used for fish and wildlife purposes; one-fifth of this land is managed by other federal and state agencies in cooperation with the Corps.

*"Tanks to Reef" Project in Mobile District: The Corps used surplus tanks to help create a 1,000-square-mile artificial reef zone near Mobile, Ala., 1994.*

The intense interest of the Corps in fish and wildlife management derives in part from the program's value to the recreational functions at 456 Corps water resource projects covering an aggregate of more than 11.5 million acres. Nearly 400 million visitors annually enjoy fishing,

*Fish viewing window at Lake Washington Ship Canal, Wash., 1990*

*Fish ladder on Little Goose Dam, Wash.*

hunting, swimming, and other water-related sports at Corps recreation areas.

Through its floodplain management program begun in 1960, the Corps provides technical services and planning guidance for many local agencies and groups to encourage prudent use of floodplains. At the request of state or local agencies, the Corps identifies flood hazard potentials, establishes standard project floods (the flow that can be expected under conditions of maximal severity), devises flood frequency curves, and maps the floodplains. The resulting information is used by the local agencies to regulate floodplain development—even to the extent of evacuating flood-prone areas and converting them to recreation parks or fish and wildlife habitats.

*Aerial view of Mallard Island, Minn., in the Mississippi River, part of the Weaver Bottoms Habitat Restoration Project*

# Saving the Salmon

Long before the construction of the famous fish ladders at Bonneville Dam, a U.S. Army engineer warned that the Columbia River salmon required protection. Major William A. Jones, an experienced engineer and explorer who discovered Togwotee Pass through the Wind River Mountains, observed over time the impediments faced by salmon in their efforts to spawn.

While serving as Portland District Engineer, Major Jones wrote his *Report on the Salmon Fisheries of the Columbia River,* published in 1888. Stunned at the maze of nets, traps, and fish wheels that clogged the Columbia near places like Astoria, he concluded that it was "a sort of miracle that any fish escape to go up the river."

Jones proposed means for mitigating the threat to the fisheries. Along with continuing the practice of closing the river to fishing at regular intervals, he recommended an increase in the number of hatcheries and uniformity between the fish laws of Oregon and Washington.

Major Jones had recognized the threat to the survival of the salmon fisheries many years before the general public would become aware of the problem. His suggestions were later adopted, but long after he first proposed them.

*Fish ladder and the Visitors Center at Bonneville Dam, Ore.*

*Coho fingerlings at Bonneville Dam*

# Work in the District of Columbia

U.S. Army engineers contributed to both the planning and construction of the nation's capital. From early bridges to the modern subway system, Corps officers and civilians helped plan and construct Washington's transportation system, city monuments, and public buildings. Parks, water supply and sewage systems, flood control structures, and public health measures in the city were or still are the engineers' responsibility. U.S. Army engineers served as administrators as well as construction experts. Their influence and responsibilities declined only as civilian agencies assumed control of certain activities and home-rule movements lessened

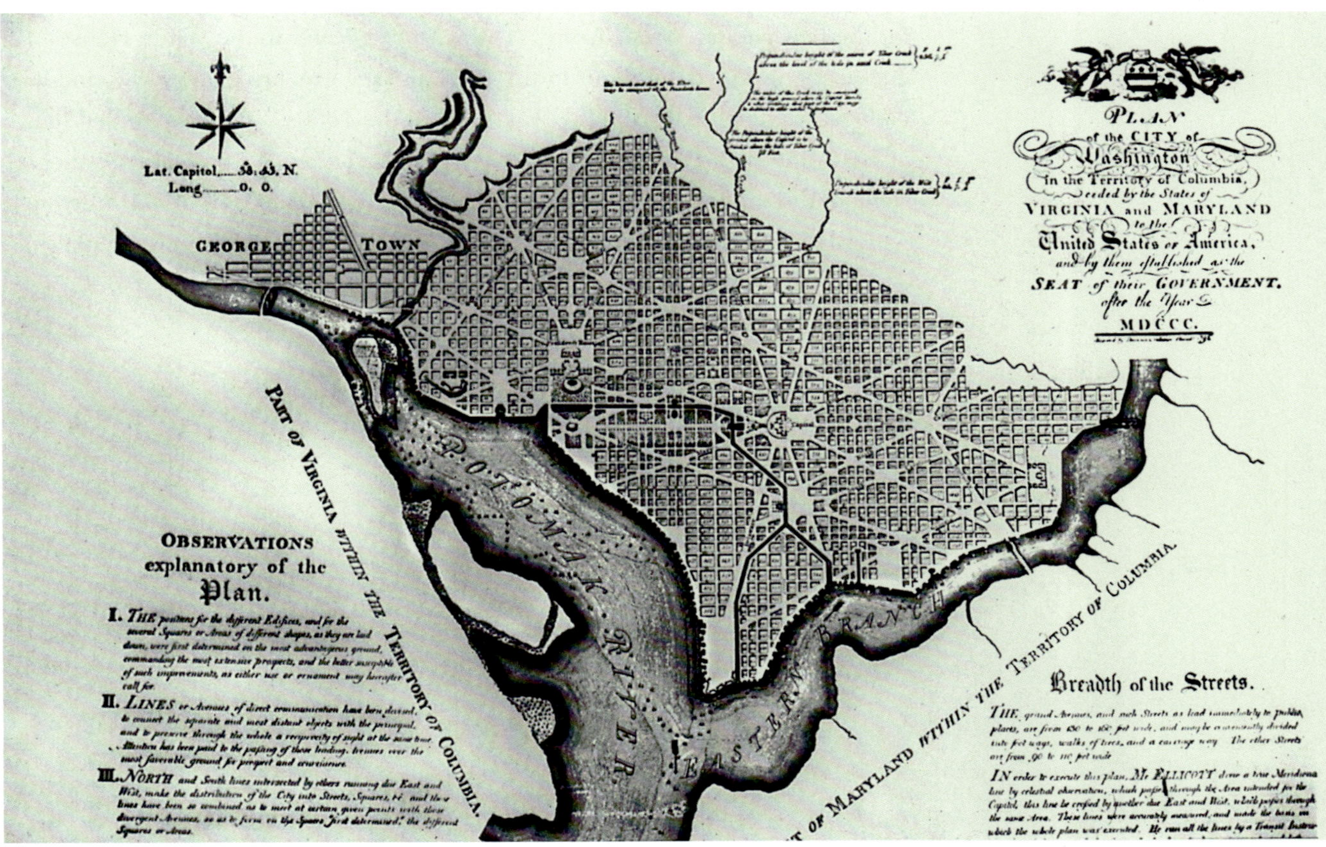


Andrew Ellicott's drawing of L'Enfant's plan of Washington, D.C., 1792

National Archives

*Arch of the Cabin John Bridge under construction, August 12, 1861*

federal responsibility for public works in Washington.

In 1791, former Continental Army engineer Pierre Charles L'Enfant designed the master plan for the new capital. Other Army engineers designed and built fortifications for the city. During the War of 1812, the British army destroyed some of those defenses as well as the partially built Capitol Building; Chief Engineer Joseph G. Swift helped rebuild the Capitol. In 1822, Major Isaac Roberdeau, a topographical engineer, supervised the installation of cast-iron pipes to bring spring water to the White House and the executive offices around it. In the 1850s, Congress funded the construction of a permanent water supply for the cities of Washington and Georgetown. Eventually placed

*U.S. Capitol dome under construction, December 31, 1857*

under the supervision of engineer Lieutenant Montgomery C. Meigs, the project evolved into what is today the Washington Aqueduct Division of the U.S. Army Engineer District, Baltimore. Lieutenant Meigs's plans included construction of two bridges to carry traffic as well as water pipes—one over Cabin John Creek and one over Rock Creek. Both bridges were engineering feats of their time, and the Cabin John Bridge remains in use. For forty years, the Cabin John Bridge (begun in 1857 and completed in 1864) held the record for having the longest masonry arch in the world.

Lieutenant Meigs and other engineer officers also reconstructed the U.S. Capitol, fireproofed the Smithsonian Institution, and rebuilt or repaired bridges and streets throughout the city. Using new techniques, Meigs provided the first adequate heating and ventilation system for the home of Congress. As construction of the two new wings of the Capitol progressed, the old dome began to look disproportionately small; a new Capitol dome was designed, consisting of cast and wrought iron and weighing almost 4,500 tons. Work on the dome continued during the Civil War.

After the Civil War, Corps officers and civilians designed and built many of the monuments and public buildings that decorate Washington today. At the request of the Senate, Major Nathaniel Michler surveyed sites for a new park and a new location for the White House. His praise drew attention to Rock Creek Valley. Later, the Chief of Engineers, Brigadier General Thomas L. Casey, and other officers worked for and supervised the development of that large, urban park.

Congress continued to institutionalize the Corps' role in the District of Columbia. In 1867, the legislators removed control of many public buildings from civilian hands and gave it to what became the Office of Public Buildings and Grounds under the Chief of Engineers. In 1878, Congress replaced Washington's elected government with a three-man commission. A U.S. Army engineer, holding the title of Engineer Commissioner of the District of Columbia, served on that governing board with responsibility for the city's urban infrastructure.

Meanwhile, other engineer work in Washington grew to such an extent that in 1874, the Chief of Engineers, Brigadier General Andrew A. Humphreys, established the United States Engineer Officer, Washington, under the civilian engineer, Sylvanus T. Abert, to carry out navigation improvements on the Potomac River and its tributaries.

Two years later, Congress asked the Corps to complete the Wash-

*"572 feet high—Setting the Cap-Stone on the Washington Monument—From a sketch on the spot by S. H. Nealy."* Harper's Weekly, *December 20, 1884. Col. Thomas Lincoln Casey is to the right of the capstone and Bernard R. Green, his chief civilian assistant, far left.*

ington Monument, left partially built by its bankrupt sponsors. Then Lieutenant Colonel Thomas Casey and his assistant, Bernard Green, corrected major problems with its foundation, redesigned it, and supervised its completion. The construction culminated in December 1884 with placing on its tip a pyramid of 100 ounces of aluminum, the largest piece of the new metal yet cast in the Western Hemisphere. Casey and Green went on to help design and supervise the construction of the State, War and Navy Building next to the White House. It is now the Eisenhower Executive Office Building. The two men also helped design and construct the Library of Congress.

In 1883, Brigadier General Meigs came out of retirement to build the Pension Building. Designed to house the offices providing pensions to war veterans, the building is so attractive that it is sometimes used for inaugural activities and is the new home of the National Building Museum.

Between the 1880s and the 1920s, Corps dredge-and-fill operations not only protected Washington from Potomac and Anacostia river floods, but also created waterfront park land. Potomac Park, the Washington Channel with its adjacent recreation areas, and the land for the Lincoln and Jefferson Memorials are all products of this river improvement and reclamation work. The attractive tidal basin in front of the Jefferson Memorial that automatically changes the water in the Washington Channel with the tidal flow is another product of this work.

Meanwhile, Lieutenant Colonel William W. Harts of the Office of

*State, War and Navy Building under construction, 1886. Army engineers completed most of the building that is now known as the Eisenhower Executive Office Building.*

*National Archives*

Buildings and Grounds accelerated the development of Rock Creek Park into a major resource for urban recreation and beauty. Lieutenant Colonel Harts also oversaw the construction of three other important memorials. In 1913, he directed the start of work on the new headquarters of the American Red Cross. The following year, he initiated construc-

*Library of Congress under construction, November 8, 1892*

*Library of Congress*

*Hoisting a twenty-ton lintel at the Army War College building at Washington Barracks (later Fort McNair) March 7, 1906*
*National Defense University*

tion on both the Lincoln Memorial and the Arlington Memorial Amphitheater and Chapel.

The Corps also built or supervised the construction of practical and attractive buildings to house the Government Printing Office and the Army War College at Fort McNair. In addition, Army engineers managed the construction of numerous bridges including the Arlington Memorial Bridge and the Francis Scott Key Bridge.

The George Washington Memorial Parkway, the Pentagon, and Ronald Reagan Washington National Airport began as pre-World War II construction projects of the Corps of Engineers. After World War II, the Corps was involved in the

*Lincoln Memorial under construction, July 1916*
*National Archives*

*Red Cross Building built as a memorial to women in the Civil War, under construction, 1916*

complete gutting and rebuilding of the inside of the White House, expanding the water supply for the District of Columbia, and planning for housing and transportation.

U. S. Grant III, grandson of the president, and other officers served on the planning boards that oversaw growth in the Washington metropolitan area. Gradually, civilian agencies, such as the National Park Service and the home rule municipal government of D.C., began to assume responsibility for developing the memorial buildings, streets, sewage systems, and parks that the Corps had once handled.

However, the Washington Aqueduct remains a special responsibility of the U.S. Army Engineer District, Baltimore, and the district continues to carry out civil works and military projects in the National Capital area.

*The Washington Engineer District built the reinforced concrete Francis Scott Key Bridge over the Potomac River from 1917 to 1923. Just above the arches of the Key Bridge still under construction, the old Aqueduct Bridge, completed in 1843 and rebuilt in the 1880s by Army engineers, was dismantled after the new bridge opened. Georgetown and Georgetown University are shown on the right.*

*National Archives*

*The Office of Public Buildings and Grounds built Arlington Memorial Bridge linking the Lincoln Memorial to Arlington National Cemetery and Custis-Lee Mansion between 1925 and 1932.*

*Library of Congress*

*The Corps of Engineers managed construction of the Pentagon, designed to consolidate most of the War and Navy departments' offices in Washington, and completed it in a remarkable sixteen months between September 1941 and January 1943.*

*A retired engineer officer managed the reconstruction of the White House from 1948 to1952 during which time the building was stripped to its bare walls. In May 1950 this bulldozer was digging more basement space for the many offices and other facilities added to the building.*
*National Park Service*

*National Airport under construction, July 1, 1940*

*The Baltimore Engineer District carried on the tradition of Corps of Engineers' work in the national capital area with a wide variety of military construction and civil works projects. One example is construction of the Korean War Veterans Memorial near the Lincoln Memorial (barely visible in the background). The Korean War Veterans Memorial is shown here nearing completion in April 1995.*

# Army Engineers and the District of Columbia Parks

One of the most beautiful areas in the Nation's Capital is Rock Creek Valley, which runs from north to south through the entire District of Columbia. In 1867, Major Nathaniel Michler, the first U.S. Army engineer to head the federal government's Office of Public Buildings and Grounds, proposed the valley as a new site for the White House.

The suggestion touched off great interest in the valley. Praising the region's "primeval forest and cultivated fields, its rocks clothed with rich ferns and mosses, its repose and tranquility, its light and shade," he saw it as a potential refuge for the president from the malarial river front and an unsightly marsh known as the Potomac Flats.

Although the White House was not relocated to Rock Creek Valley, development of the area into what became Rock Creek Park began under one of Major Michler's successors, Colonel Theodore A. Bingham. Bingham believed that the park would provide fresh air and places of recreation for crowded city dwellers and serve as an "emerald setting for the beautiful city." Other engineers shared his vision, and Frederick Law Olmsted, Jr., was hired to create the basic plan of the park and construct the parkways that would link the green areas together. Captain Lansing H. Beach would lend his name to the road he constructed that traverses the length of the park.

U.S. Army engineers also transformed the unsightly Potomac Flats. Beginning in the 1880s, the Corps dredged the river channel and dumped the material onto the flats to create new land to the south and west of the National Mall. In 1897, Congress dedicated some 638 acres of this reclaimed land and directed that it be "forever held and used as a park for the recreation and pleasure of the people." Col. Bingham personally provided Potomac Park with gardens and athletic fields. The southernmost tip of the park became known as Hains Point after engineer Brigadier General Peter C. Hains.

The Corps constructed the Tidal Basin to flush the Potomac River and help prevent pollution. This area became the center of a still-famous location of natural beauty when the U.S. Army Corps of Engineers directed the planting of donated Japanese cherry blossom trees around the basin.

Maj. Gen. Lansing H. Beach as Chief of Engineers, 1920–1924

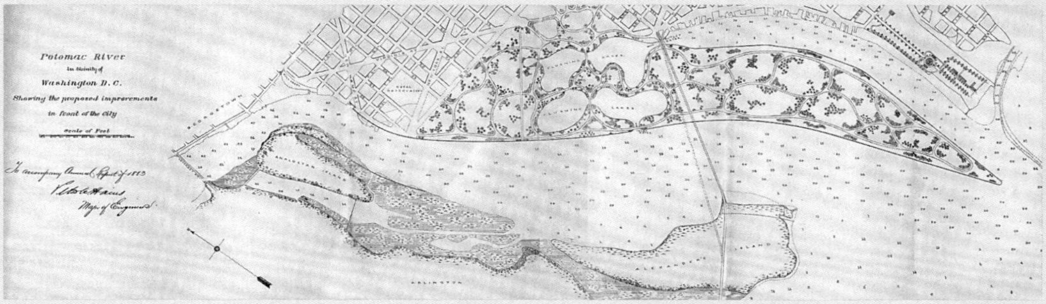


*Plan for Improvement of Potomac Flats by Major Peter C. Hains, 1882*

# 3 Strategic Role in War and Peace

*Interior view of the casemates at Fort Jefferson, Fla.*

# Coast Defense

When the American Revolution began in 1775, numerous coastal fortifications already existed along the Atlantic Coast to protect communities from pirate incursions and enemy raids. The British Royal Engineers, as well as individual colonies and local communities, built these structures, which varied from crude earthen and wooden batteries to strong masonry forts.

During the War for Independence, the combatants rehabilitated many of the existing coastal fortifications and constructed new ones. The small body of Continental Army Engineers accomplished some of the work. When the war ended, the new country abandoned these works, deciding that the local militia could man them if necessary.

A decade later, in 1794, the United States, fearing attacks from other nations, began a construction program to provide fortifications for the protection of the major harbors and northern frontiers of the country. This program and another on the eve of the War of 1812 made only modest progress in strengthening

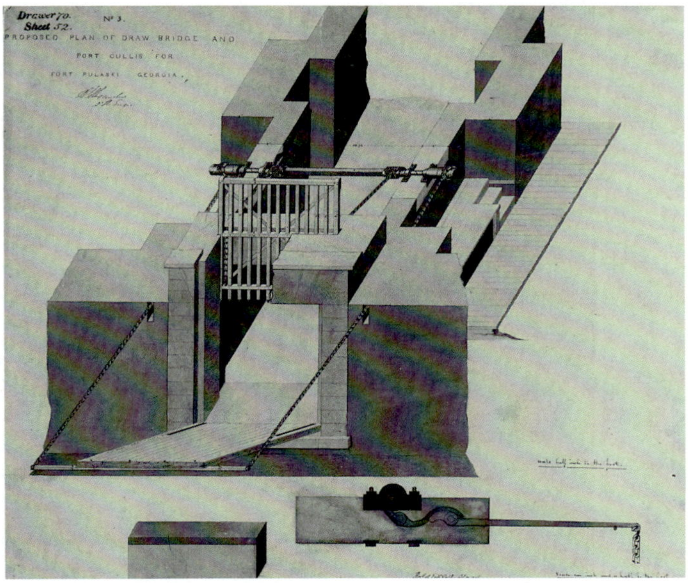


*Drawbridge plans for Fort Pulaski, Ga., c. 1846*

*National Archives*

*Fort Jefferson in the Dry Tortugas was designated a National Monument in 1935.*

*National Oceanic and Atmospheric Administration*

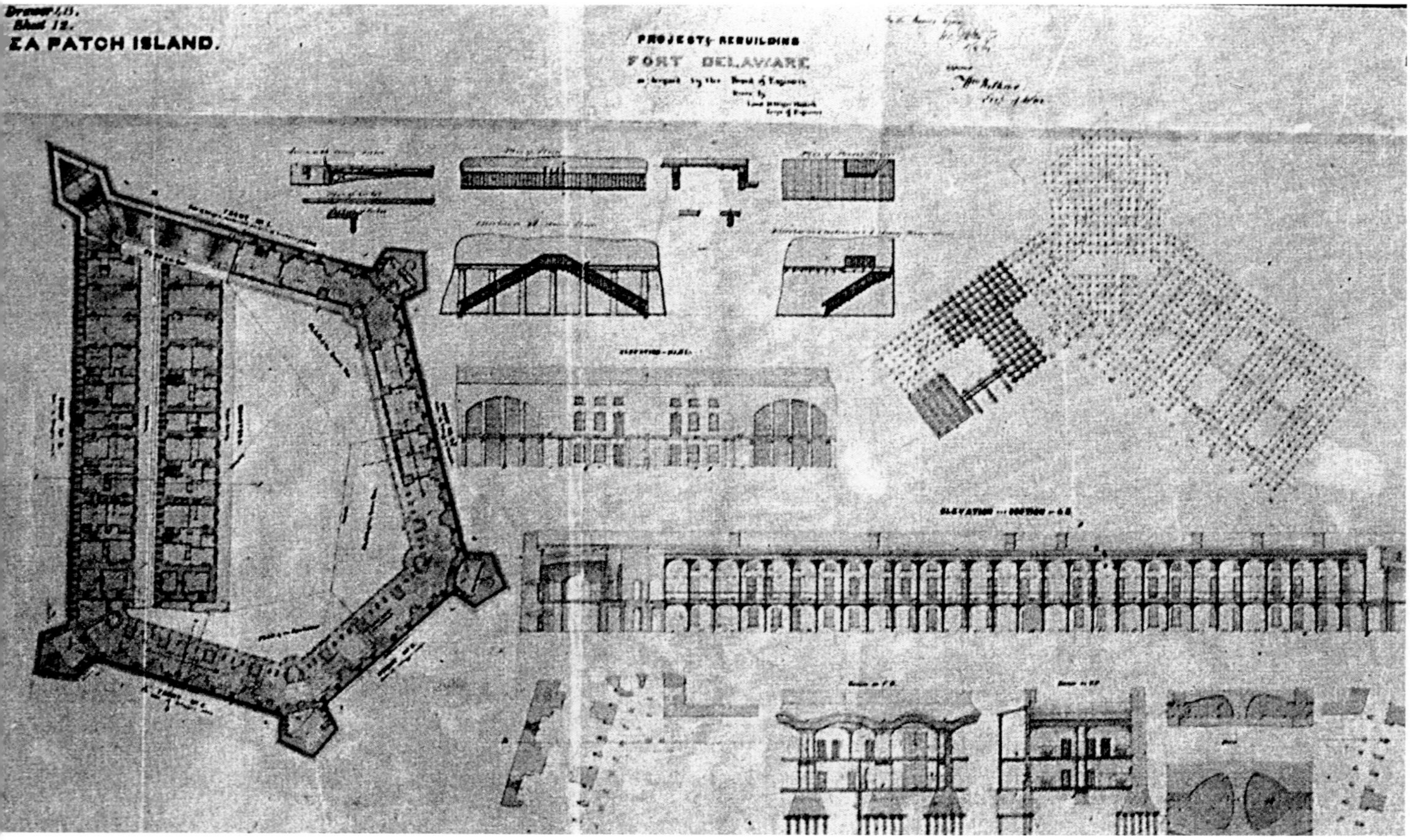


Plan for the third fort started on Pea Patch Island in the middle of the Delaware River. This design for Fort Delaware from 1839 constructed the fort on a wooden grillage depicted in the upper right.

Brig. Gen. Seth Eastman's oil painting of Fort Sumter, as it looked before the Civil War.

the country's coastal defenses; however, the burning of the Capitol and White House and attacks on other coastal areas led to a more concerted post–War of 1812 effort to build substantial and sophisticated fortifications. Initially Army engineers followed the prevailing design principles taken from the famous seventeenth century French engineer, Vauban, but gradually the engineers adopted a variety of designs, some influenced by the most sophisticated and novel

*(top) Ten-inch disappearing gun of the Endicott system in the loading position at Sandy Hook, N.J.*
*National Archives*

*(center) Ten-inch gun in firing position*
*National Archives*

*(bottom) Mortar battery at Sandy Hook, N.J.*
*National Archives*

European principles. Fort Monroe in Virginia, Fort Adams in Rhode Island, and Fort Washington in Maryland exhibit traditional influence, while Fort Delaware in Delaware and Fort Point in California reflect newer concepts.

Although generally ungarrisoned, the country's coastal fortifications were a deterrent to foreign attack until the Civil War, when newly developed weapons and ships rendered them obsolete. Heavy rifled artillery, both land and naval, demolished brick, stone, and masonry fortifications like Fort Sumter, South Carolina, and Fort Pulaski, Georgia. As a result, both Union and Confederate engineers began erecting earth and wood coastal forts and batteries that were much more resilient to artillery fire.

For two decades after the Civil War, America's coast defenses received little attention, but by the mid-1880s the sad state of the defenses led to the appointment of a board, named the Endicott Board, after the Secretary of War. In 1886 the board recommended an ambitious program that was gradually scaled back. Even so, the new defenses incorporated the latest technology including

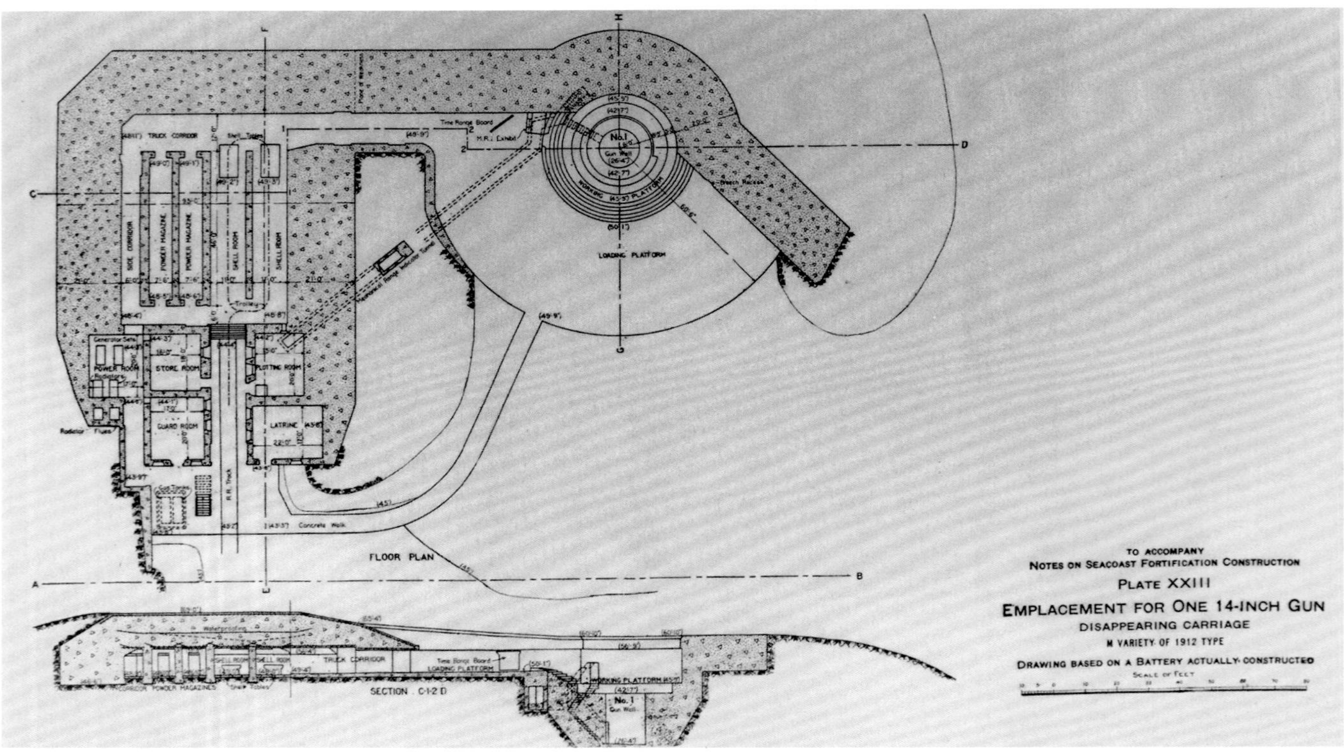


*Plan of an emplacement for a fourteen-inch disappearing carriage gun from Col. Eben E. Winslow, "Notes on Seacoast Fortification Construction," published in 1920 as an engineer* Occasional Paper *for instructional use at the Engineer School, Washington, D.C.*

breach loading, disappearing guns arranged in dispersed batteries; heavy mortars whose shells were to penetrate the lightly armored decks of ships; and mines to obstruct waterways. Army engineers sometimes placed the batteries inside or in the immediate vicinity of old coastal forts; they purchased new land for others. With the acquisition of new territories at the end of the century, the engineers began erecting batteries in Hawaii, Panama, and the Philippines. As artillery improved, the Corps constructed new batteries for bigger and more effective guns.

After World War II, new weapons—airplanes and missiles—rendered the coastal batteries obsolete. By 1950, the U.S. Army ceased using them for their original purpose. Today, the remnants of these batteries dot the coasts and from a distance often resemble concrete bunkers.

In conjunction with its fortification and battery construction programs, the U.S. Army Corps of Engineers had other coastal defense responsibilities. In the nineteenth century, the Corps placed obstructions in the bays, rivers, and harbors along the coasts.

*Fort Moultrie, S.C., in camouflage during World War II*

These obstructions—from chains to submarine mines—were intended to slow down or halt enemy vessels. Although the Coast Artillery Corps took over responsibility for submarine mines in 1901, the U.S. Army Corps of Engineers continued to build casemates, storehouses, loading rooms, and other structures for the mine defenses. The Corps also developed a protective concealment program for coast defenses that evolved into the elaborate camouflage nets and paints used during World War II.

# Engineering for Posterity

Sound workmanship is a long-standing tradition within the U.S. Army Corps of Engineers and is exemplified in an early project the Corps undertook near the Nation's capital—Fort Washington on the Potomac.

Pierre L'Enfant had only just begun construction of a new fort on the site of an earlier one destroyed during the War of 1812 when he left the project. When construction on the fort resumed in 1815, Colonel Joseph G. Swift instructed Lieutenant Colonel Walker K. Armistead, "Let us have it done for posterity, or not at all." Lieutenant Colonel Armistead replied that he would build a fort "exceedingly strong, of the most durable materials, and executed in the best manner." History has proven that the U.S. Army Corps of Engineers succeeded in that mission.

At the outset of the Civil War, Fort Washington was the only defense for Washington, D.C. The U.S. Army continued to occupy the fort as a major defensive post until the eve of World War I. It subsequently served as home to ceremonial units, an officer training school, and the site of a Veterans Administration hospital. In 1946, the fort was turned over to the Department of the Interior and became a national park. The old fort, its fortifications remaining in original form, still stands as a major landmark and a testament to the technical expertise of the U.S. Army Corps of Engineers.

*Aerial view of Fort Washington, now part of the National Park Service system*

*1823 plan of Fort Washington, Md.*

*National Archives*

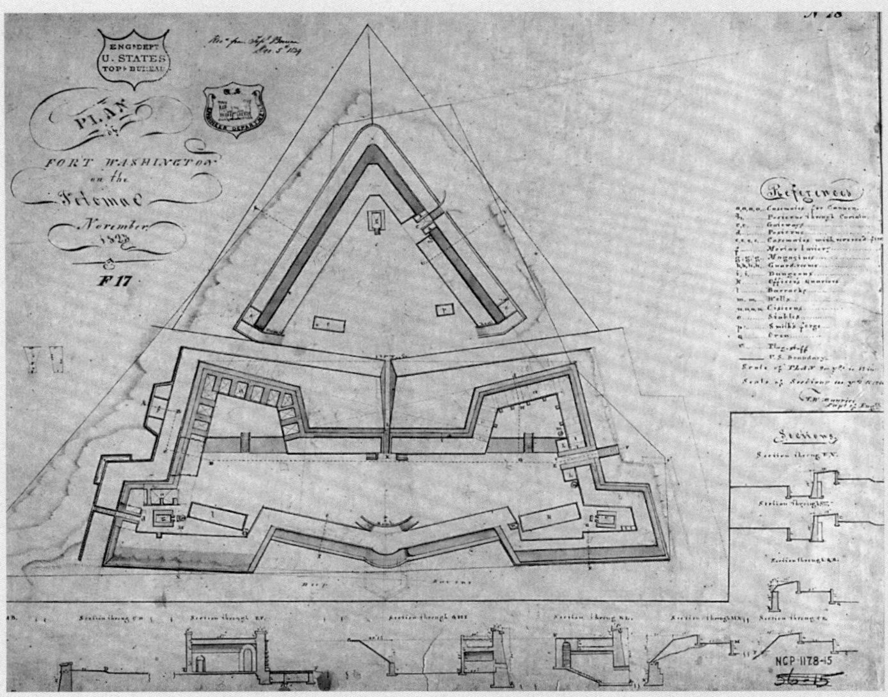

Civil War ponton bridge
National Archives

# Combat Operations, 1846–1916

As the United States developed and expanded throughout the balance of the nineteenth and into the early twentieth centuries, the U.S. Army Corps of Engineers played a key role during times of war. Engineer troops have performed heroically in support of the war-fighting mission, and as a consequence the Corps established a history of wartime service that truly demonstrated the value of military engineering to success on the battlefield.

## The Mexican War

On May 15, 1846, soon after the Mexican War began, Congress authorized the War Department to raise a company of engineers. This unit, the first regular U.S. Army engineer company fielded, acted as

*Siege of Monterey, July 1846*
*National Archives*

Ponton bridge, held in place by ships, across the James River, June 1864.

Topographical engineers at Camp Winfield Scott near Yorktown, Va., May 1862, before the two corps were reunited in 1863.

Rappahannock River under devastating fire from Confederate sharpshooters. In June 1864, Army of the Potomac engineer troops constructed a 2,170-foot ponton bridge across the James River, one of the longest floating bridges ever constructed in modern times.

When the Civil War began, two engineer corps existed in the Union Army: the Topographical Engineers and the Corps of Engineers. But the exigencies of the war required stricter coordination of engineer activities. Therefore in 1863, the War Department integrated the smaller Corps of Topographical Engineers into the Corps of Engineers under the command of the Chief Engineer.

Pre-war engineers McClellan, Halleck, George G. Meade, William S.

Rosecrans, William B. Franklin, Gouverneur K. Warren, James B. McPherson, and Andrew A. Humphreys did not serve on the battlefields as engineers. Instead they were promoted to general officers commanding combined troops. Likewise, Montgomery C. Meigs became the quartermaster general of the Union Army and furnished the required support and supplies to the troops in the field. By the end of the war, James H. Wilson was a cavalry general.

Their engineering expertise allowed these former Corps officers to excel. As the Battle of Gettysburg unfolded during the summer of 1863,

*Company A, Battalion of U.S. Engineers, 1865*

*1st Michigan Engineers and Mechanics Regiment built this four-tiered, 780-foot railroad trestle bridge, Whiteside, Tenn., 1864.*

*National Archives*

Warren used the talent for assessing terrain he had gained from earlier engineering assignments to discern a weakness in the Union lines along the position known as Little Round Top. He quickly strengthened that position and thereby foiled a key part of the Confederate battle plan.

Other able officers—like Henry Brewerton, John G. Barnard, and Nathaniel Michler—were engineers throughout the war. These men conducted surveys and reconnaissances to provide intelligence reports and maps, directed siege operations, and oversaw the operations of engineer troops. Competent volunteer engineer officers, like William G. Margedant, who developed a process for duplicating maps in the field, also greatly aided the Union war effort.

Three young engineer lieutenants —William H. H. Benyaurd, John M. Wilson, and George L. Gillespie—received Medals of Honor for gallantry under fire, and the latter two concluded their U.S. Army careers as Chief of Engineers. Lieutenant Wilson received the Medal of Honor for his actions at the Battle of Malvern Hill in 1862; Lieutenant Gillespie received the Medal of Honor for actions at the Battle of Cold Harbor; and Lieutenant Benyaurd won his medal at the Battle of Five Forks, Virginia.

The Confederacy gladly accepted the services of fifteen engineer officers who had resigned their commissions in the U.S. Army. Former engineer officers such as Lee, Beauregard, and Joseph E. Johnston became Confederate Army commanders. Edward P. Alexander was the Confederate artillery commander in the Army of Northern Virginia. To

*Students at Willets Point building a ponton bridge, 1889*

accomplish the necessary engineer work, the Confederacy commissioned many former civilians and raised engineer and pioneer units.

## Post-Civil War Period

Between the end of the Civil War and the outbreak of the Spanish-American War, engineer combat experience was minimal. Most engineer officers returned to civil works or fortification construction duty, although they attempted to stay abreast of new military engineering methods and innovations.

Soon after the Civil War ended, Congress abolished the U.S. Army Corps of Engineers' supervision of the U.S. Army Military Academy at West Point, New York. Therefore the Corps, unofficially at first, established an Engineer School at Fort Totten at Willets Point in New York Harbor in 1866. The school's staff instructed students—both officers and enlisted men—in civil and military engineering and provided practical training in mapping, military photography, and laying submarine mines and bridges, both ponton and trestle. In addition to teaching, the staff, especially Superintendent Henry L. Abbot, experimented with and developed new equipment.

Some engineer officers served with the "Indian-fighting army" on the western frontier. A few, like William Ludlow, accompanied the

*Underwater mine testing at the Engineer School, Willets Point, N.Y.*

troops on reconnaissance missions and scouting expeditions. Generally, though, these officers' main duties were surveying and mapping.

Other officers, such as Barton S. Alexander, Cyrus B. Comstock, Peter S. Michie, John M. Wilson, William Craighill, and William E. Merrill, traveled abroad, sometimes as military attachés. Often they had the chance to observe foreign engineer troops' equipment and techniques. A few, including Francis V. Greene, actually witnessed engineer operations in battle.

The War Department created a fifth regular Army company of engineers in December 1865. Between the Civil War and the Spanish-American War, the five companies of the battalion, usually understrength, performed a range of duties, from serving at engineer depots in New York Harbor, St. Louis, and San Francisco to riot control during the 1877 railroad strikes. Individual engineer soldiers assisted at numerous civil works and fortification sites throughout the country.

## The Spanish-American War and the Philippine-American War

In 1898, the United States went to war with Spain, and the engineers provided extensive combat support. In the far-flung theaters of the war, from Cuba and Puerto Rico to the Philippines, the engineers aided the U.S. Army by erecting landing piers, constructing bridges, building and maintaining roads, laying mines offshore, and repairing and operating railroads. Young but capable lieutenants like Lytle Brown, Eben E. Winslow, and William D. Connor led engineer detachments on dangerous reconnaissance missions, sometimes in the midst of combat. Volunteer engineer units, often commanded by regular U.S. Army officers, also

*Engineers' train in the Philippines, 1899*

served in the war. Former engineer officers, such as Francis V. Greene and William Ludlow, were brigade and higher unit commanders.

Following the Spanish-American War, an insurrection broke out in the Philippines. Companies A and B of the Engineer Battalion served in the initial stages of the conflict. The insurrectionists' guerrilla warfare tactics necessitated rapid movements by the U.S. Army. Thus, engineer detachments, commanded by William Sibert, John Biddle, John C. Oakes, and Harley B. Ferguson, among others, had to repair roads, build bridges, and perform reconnaissance rapidly over difficult jungle and mountain terrain. Frequently, the engineer troops, who carried rifles as well as picks and axes, joined the infantry in fighting off an attack before completing work on a road or bridge. The requirements of combat, especially in the Philippines, influenced the 1901 reorganization of the engineers into three battalions of four companies each. Although the fighting subsided in the Philippines in the early twentieth century, it did not cease, and engineer troops served in the islands, often in combat, for many years afterward.

*Company H, 1st Provisional Battalion of Engineers, near Guánica Bay, where U.S. forces landed on the southern shore of Puerto Rico, July 1898.*

## The Mexican Punitive Expedition

In 1916, the U.S. Army Corps of Engineers formed three regiments of six companies each from the battalions. In the same year, the United States launched a punitive expedition to Mexico to chastise the "bandits" under Pancho Villa, who had raided American territory. The use of cars and supply trucks required better roads and bridges than ever before. Lytle Brown, now a major, was one of many engineer officers who served in Mexico. These officers gained experience that became especially valuable after April 1917, when the United States entered World War I.

*The sinking of the* U.S.S. Maine *in 1898 inflamed public opinion and pushed the U.S. into war with Spain. After the Spanish-American War, the Corps of Engineers built caissons in 1911 around the* Maine *in Havana harbor and pumped out water so the ship could be examined before it was towed to deep water and sunk in its final resting place. The USACE Museum Collection has the* Maine's *ship's wheel that the Corps received in appreciation for its work in raising the famous ship.*

# We Don't Surrender Much!

At the end of 1862, Colonel William P. Innes and 391 men of the 1st Michigan Engineers were repairing roads and railroads at the rear of the Union Army near Murfreesboro (Stone's River), Tennessee, when a Confederate cavalry division, commanded by General Joseph Wheeler, flanked the Union Army to strike hard at supply trains on the way from Nashville to Murfreesboro. The surprise attack left Colonel Innes and the engineers without time to escape the gray-clad troopers, and Innes rushed his unit up a nearby hill.

From the top of the hill, Colonel Innes could see the advancing Confederate columns and realized he had no time to entrench his position. But the hill was covered with clumps of red cedar trees, and Innes quickly decided to use this resource. He sent the engineers scrambling around the hill, slashing down the small trees to open a field of fire and piling the cedars in a waist-high circle around the crest of the hill.

Confederates, in greatly superior force, soon surrounded the hill. An officer under a flag of truce advanced to demand surrender from the engineer detachment and was surprised by Colonel Innes' acerbic reply: "Tell General Wheeler I'll see him damned first." Innes continued, "We don't surrender much. Let him take us."

Confederate cavalry soldiers swept up the hill toward the engineers' position, but a volley of Union fire hurled them back pell-mell. The Confederates then unlimbered field artillery and began pounding the hill. The engineers scraped shallow foxholes and held their place. A second cavalry assault followed, and then a third. In all, the cavalry made seven attempts to take the hill, yet the engineers stood their ground until the Confederates concluded the effort was not worth the cost. The engineers suffered eleven casualties, the Confederates nearly fifty.

*1st Michigan Engineers and Mechanics Regiment, Company D, on dress parade*

*West chamber of the Gatun Upper Locks, March 1912*
*U.S. Military Academy Library*

# The Panama Canal

In the early morning of May 4, 1904, a young lieutenant from the U.S. Army Corps of Engineers crisply walked into the old French Hotel in Panama City. He exchanged brief greetings with officials of the new French Panama Canal Company. The new company, which had succeeded Ferdinand de Lesseps' bankrupt enterprise in 1894, had been no more successful than its predecessor in the effort to build a canal across the Isthmus of Panama connecting the Pacific and Atlantic Oceans. Its workers ravaged by yellow fever and malaria and its equipment in disrepair, the company was ready to sell all of its assets to the U.S. government for $40 million. The lieutenant carefully read the document of transfer. Then, following the directions of the U.S. Secretary of War, he signed his name to the receipt: "Mark Brooke, 2nd Lieutenant, Corps of Engineers." The long years of the French effort to construct an isthmian canal were over. The American attempt was about to begin.

Building the Panama Canal required the assistance of the foremost engineers of the day. Major William M. Black, who later became Chief of Engineers, supervised early engineering activities at the canal. John F. Wallace, the first civilian chief engineer on the project, brought railroad construction and operations expertise to the isthmus. His successor, John F. Stevens, continued his endeavors and established the basic plan for the construction of the canal. Stevens resigned, however, in 1907 when he was severely criticized in the United States.

Frustrated by his inability to find a civilian willing to see the project through to completion, President Theodore Roosevelt turned for help to the U.S. Army Corps of Engineers. "We can't build the canal with a new chief engineer every year," he said. "Now I'm going to give it to the Army and to someone who can't quit." He requested the Panama Canal Commission to appoint engineer officer Lieutenant Colonel George W. Goethals as Chief Engineer and commission chairman. Engineer officers Major William L. Sibert and Major David D. Gaillard, both West Point graduates like Lieutenant Colonel Goethals, also served on the commission. All

*Lt. Col. George W. Goethals*

*Maj. David D. Gaillard*

Construction of Pedro Miguel Lock, 1911

U.S. Military Academy Library

three men received several promotions during the time they worked on the canal.

Within a year, Lieutenant Colonel Goethals reorganized canal operations into three geographical divisions. Major Sibert took charge of the Atlantic Division, and Major Gaillard took the Central Division. To head the Pacific Division, Goethals selected Sydney B. Williamson, a civilian engineer who had won Goethals's respect when the two had worked together earlier at Muscle Shoals. The civilian engineers under Williamson engaged in a spirited competition with the military engineers. Lieutenant Colonel Goethals encouraged this competition to achieve maximum economy while speeding construction. Rear Admiral Harry H. Rousseau, chief of the Bureau of Yards and Docks of the Navy, assumed responsibility for the design and construction of terminals, wharves, docks, warehouses, machine shops, and coaling stations. Civilian engineer Ralph Budd directed the relocation of the Panama Railroad from 1907 until 1909, when he was succeeded by Lieutenant Frederick Mears, an Army cavalry officer.

Miraflores Locks under construction, August 1912

U.S. Military Academy Library

*Steam shovel at work in the Culebra Cut; a lithograph in a series on the Panama Canal by Joseph Pennell, noted American artist and illustrator.*

*USACE Museum Collection*

*A rail line assisted the canal's construction.*

In the 1880s, the French had learned, after several years of effort, that a sea-level canal across Panama was an impossibility. Locks were absolutely necessary. Benefiting from French experience, the Americans never seriously considered anything other than a canal using locks. They erected a monumental dam across the Chagres River, thereby creating Lake Gatun. At each end of the lake, the engineers constructed locks. The Gatun Locks lead to the Atlantic. The Pedro Miguel Locks lead to Miraflores Lake and, farther on, Miraflores Locks. From these locks, ships travel on to the Pacific.

Major Gaillard directed the huge engineering task of completing the Culebra Cut through the continental divide, which required the excavation of 96 million cubic yards of rock and dirt. Spectacular landslides at the cut were the greatest engineering difficulty. The amount of earth that had to be removed was nearly double the original estimate. More than 100 steam shovels removed most of the soil, and flatcars hauled it out. Trains departed at thirteen-minute intervals to keep pace with the steam shovels.

Construction of the Panama Canal was the responsibility of the Panama Canal Commission, but having Army engineer officers supervising the project enabled problems to be resolved more easily and quickly. Engineer officers worked effectively and completed the canal well within estimates. Going beyond mere construction, they also helped

*Drilling at Culebra Cut, January 1912*
*U.S. Military Academy Library*

eradicate disease and vastly improved sanitation in the areas adjoining the canal. The organization, administration, and implementation of this massive building effort remain a model for subsequent large-scale construction projects.

*Culebra Cut*
*U.S. Military Academy Library*

The Panama Canal opened ahead of schedule on August 15, 1914. The total excavation for the channel exceeded 200 million cubic yards of earth, of which almost half was taken from the Culebra Cut, later renamed Gaillard Cut in honor of the officer who conquered it. Tragically, Lieutenant Colonel Gaillard died of a brain tumor in 1913 without seeing the canal's completion.

U.S. Army engineers retained a unique relationship with the Panama Canal after the canal was opened. Engineer officers traditionally served as the governor and lieutenant governor of the Panama Canal Zone. The governor also served as president of the Panama Canal Company, which was actually responsible for canal operations. Goethals himself was the first civil governor of the Canal Zone and received a promotion to major general during his tenure. The last military governor of the Canal Zone was Major General Harold R. Parfitt, a U.S. Army engineer officer, whose tenure ran from 1975 to 1979.

In the years immediately after the canal's completion, the U.S. Army Corps of Engineers accepted responsibility for dredging the channel, which continued to be frequently blocked by landslides. Engineers finally determined the proper incline for the banks to provide the greatest assurance against slides. In the 1920s, the Corps further strengthened the banks by developing a system of drainage control. Still later, U.S. Army engineers helped enlarge the canal. The original locks are still in use.

U.S. Army engineer officers have also periodically assisted in studies on other canal routes across Central America. Engineers conducted a survey for a route across Nicaragua in the 1930s. In the 1960s, they were heavily involved in studies on an alternate Panamanian route that would accommodate larger vessels. Although the United States turned over control of the canal to Panama on December 14, 1999, the strategic fifty-mile waterway remains a lasting testament to the skill of U.S. Army engineering.

*USS* Saratoga *in Gaillard Cut, February 1928*

*Engineers built Fort de Lesseps in 1911 to protect the canal.*
*National Archives*

# Emergency Nuclear Power for the Panama Canal

One of the most unusual ways U.S. Army engineers assisted canal operations occurred in 1968 when the Corps sent the *Sturgis,* the world's first floating nuclear power plant, to the Canal Zone to alleviate dangerous reductions of electrical power caused by necessary curtailment of operations at the Gatun Hydroelectric Station.

The weather had been so dry that there was not enough water to operate the locks as well as supply the turbines. Because of the increased traffic in the Panama Canal resulting from the Vietnam War and the closing of the Suez Canal, such vast amounts of water were required to operate the locks that the water level on Gatun Lake fell drastically during the dry season. Serviced by hyrdoelectric plants with a combined output of approximately 100 megawatts, the Canal Zone had insufficient reserve capacity to shut down its largest generator without interrupting power supply to military or civilian consumers.

In this emergency the U.S. Army Corps of Engineers dispatched the *Sturgis* to Gatun Lake. The 10-megawatt floating power plant had been designed by the Philadelphia Engineer District and christened in 1964 in memory of Lieutenant General Samuel D. Sturgis, Jr., the former Chief of Engineers who had died that year. Home port for the *Sturgis* was at Gunston Cove on the Potomac River, and its crew trained at Fort Belvoir, Virginia.

Towed to the canal, the *Sturgis* was connected to the Panama Canal Company's power grid and began producing electricity on October 5, 1968. An additional barge with greater capacity was deployed the following month to assist the mission.

The *Sturgis* fulfilled a critical power need. It also helped save more than one trillion gallons of water for lock operations that otherwise would have been used for electrical generation. The ingenuity of the U.S. Army Corps of Engineers had paid off.

*(top) U.S.S.* Sturgis, *housing the MH-1A nuclear power plant, in the Panama Canal, 1970*

*(above) Lt. Gen. Samuel D. Sturgis, Jr.*

# U.S. Army Engineers in World War I

During World War I, the U.S. Army Corps of Engineers was called upon to provide a much more diverse range of military services than ever before. Not only did the engineers provide American combat divisions with the officers and men to staff the 1,660-man engineer regiments that were part of each combat division, they also built the port facilities, roads, and railroads needed to bring essential war materiel to the front; harvested timber for military construction; employed searchlights in antiaircraft defense; organized the first U.S. Army tank units; and developed chemical warfare munitions and defensive equipment. So important were these last pursuits that, in 1918, the Army created a separate Tank Corps and a Chemical Warfare Service, the latter headed by an engineer officer.

The U.S. Army engineers who served in World War I brought with them varied amounts of military experience. Most senior engineer officers were graduates of the U.S. Military Academy and had previously served with U.S. Army units abroad, primarily in Cuba or the Philippines. A few of them had accompanied General John Pershing in his expedition to northern Mexico in 1916–1917, which had unsuccessfully attempted to capture the Mexican revolutionary Pancho Villa after his raid on Columbus, New Mexico. Some engineer commanders had been civilian engineers, members of the National Guard, or Officers Reserve Corps engineer units organized a few years before the United States' entry into the war. But most of the 240,000 engineers who served in Europe during the war had no prior military service.

The British and French governments made the arrival of American

*Company D, 11th Engineers, building a road near the Aire River*
*U.S. Military Academy Library*

engineers in France their top priority after the United States declared war on April 6, 1917. Thus, by the end of August 1917, nine newly organized engineer railway regiments, together with the engineer regiment of the 1st Division, had crossed the Atlantic and arrived in France. Several of the railway regiments were assigned to British or French military formations pending the arrival of larger numbers of American combat troops in the summer and autumn of 1918. It was while serving with the British near the village of Gouzeaucourt, southwest of Cambrai, France, on September 5, 1917, that Sergeant Matthew Calderwood and Private William Branigan of the 11th Engineers were wounded by artillery fire, becoming the first U.S. Army casualties of the war. When the Germans launched a counteroffensive in late November 1917 to regain territory they had just lost to the British near Cambrai, the men of the 11th Engineers abandoned their railway work and assisted the British with constructing new defensive positions, which stopped the German advance.

During 1918, U.S. Army engineers served in combat from the Vosges Mountains near the Swiss border north to Oudenaarde, Belgium. One battalion of the 310th Engineers served in the Murmansk area of northern Russia in a mission to assist Czech troops to rejoin the fighting on the Western Front after Bolshevik Russia had left the war in March 1918. Most of this combat service consisted of constructing

*Company E, 21st Engineers, operating a train near Menil-la-Tour, France, March 1918*

bridges, roads, and narrow-gauge railroads at or immediately behind the front, but engineer units also engaged in direct combat.

Two companies of the 6th Engineers ceased their construction of heavy steel bridges to join British and Canadian forces in frontline trenches. Together they successfully defended Amiens from a heavy German assault in March and April 1918. These two engineer companies suffered a total of 77 casualties. During June and July 1918, troops of the 2d Engineers fought as infantry in their division's bitterly contested capture of the Belleau Woods and the nearby village of Vaux in the Aisne-Marne campaign. A battalion of the 1st Engineers fought as infantry in the capture of Hill 269 in the Romagne Heights along the Hindenburg Line on October 8, 1918. It was for his action during this battle that engineer Sergeant Wilbur E. Colyer of South Ozone, New York, received

*(above) 21st Engineers maintaining a narrow gauge rail line to supply ammunition to the front, April 1918*

*(left) 107th Engineers building a bridge, Cierges, France, August 1918*

Road construction, France

During the war, U.S. Army engineers also drew and printed maps, conducted geological studies with an eye to underground water supplies, installed and operated electrical lines and mechanical equipment, and experimented with the use of tractors and trailers for hauling ponton bridging equipment in the absence of sufficient draft animals. American engineers also operated seven cement plants in France. These varied operations permitted the U.S. Army to field and support a force of nearly two million men in France within twenty months of the U.S. entry into the war.

Engineers laying foundation for barracks and hospital in France

# Maintaining High Standards: The 2d Engineers in France

The 2d Engineers had their start during the Civil War and saw action during many major battles in that conflict. The unit also participated in the Spanish-American War and the Punitive Expedition against Mexico.

During World War I, the 2d Engineer Regiment of the 2d "Indian Head" Infantry Division, commanded successively by Colonels James F. McIndoe and William A. Mitchell, was considered one of the best regiments in the American Expeditionary Forces (AEF) in France. Because of its bloody engagements at Belleau Wood, Château Thierry, Soissons, and Meuse-Argonne, the division's infantry units sustained the highest percentage of major casualties among all AEF units—its 30.38 percent casualty rate just edging the 30.08 percentage of the "Big Red 1," the 1st Infantry Division.

The 2d Engineers, moreover, stood 15th in the list of casualties with 12.73 percent, by far the highest of any U.S. Army engineer unit. The reason was simple: the trench war was preeminently an engineers' war—cutting barbed wire entanglements; putting them up; digging dugouts, machine-gun positions, and trenches; and all too often fighting as infantry.

Throughout its time in combat, the regiment maintained high morale and unexcelled performance in all its assignments. A major reason for its excellent performance was the high standards its officers and men required of themselves and each other. These standards applied throughout the regiment and were vigorously enforced.

An unnamed American general officer reinforced this assertion by noting that "the 2d Engineers is the best regiment I ever saw. . . . The regiment has assisted the artillery, has helped the tanks, built railroads, manned machine guns, and fought time after time as infantry. That regiment can do anything."

The 2d Engineers lived up to their motto, "Ardeur et Tenacite." The unit received the Croix de Guerre from the government of France.

*2d Engineer Regiment Distinctive Unit Insignia*

*Dugout entrance, Argonne, France, 1918*

*Trench warfare, France*

*Vehicles of the 3rd Armored Division cross the Seine River on an engineer-built ponton bridge, August 1944*

# Combat Engineers in World War II

As Imperial Japanese forces expanded their conquest of China and Nazi Germany gained territory in Central Europe during the late 1930s, the U.S. Army Corps of Engineers numbered fewer than 800 officers and 6,000 enlisted men in active Regular Army service. During the years since the 1922 withdrawal of U.S. Army engineer troops from Coblenz, Germany, where they had occupied territory along the Rhine River, the U.S. Army had maintained on active duty only eight or nine combat engineer regiments, two engineer squadrons, and a single topographic battalion. Furthermore, it staffed even this short troop list at only some 70 percent of authorized strength. Engineer officers thus spent most of their time during the 1920s and 1930s administering the Corps' civil works program, whose budget in 1938 was nearly four hundred times greater than its military budget.

Engineer military mobilization began in earnest in mid-1940, after the German conquest of France. During late 1940 and early 1941, the U.S. Army inducted eighteen National Guard divisions, each containing an engineer combat regiment, and their men began to undergo intensive training. The U.S. Army quickly organized engineer aviation companies and battalions to build the airfields needed to defend the Western Hemisphere.

A source relatively untapped in previous wars, African-Americans joined the U.S. Army in unprecedented numbers during 1940 and 1941. Many were assigned to engineer units. Black Soldiers, who numbered 20 percent of Corps personnel by war's end, were assigned to

*Engineers train in constructing an assault ponton bridge, Fort Belvoir, Va.*

segregated units, usually in the construction field, but they were trained by white officers such as Major (later General) Andrew Goodpaster.

Initiated well before the attack at Pearl Harbor, engineer research and development projects directed by the Engineer Board at Fort Belvoir, Virginia, were to have a significant impact upon the war. Experiments conducted during 1940 and 1941 developed a light and inexpensive pierced-steel plank mat that the U.S. Army Air Forces would widely use to provide safe, stable landing fields for American planes. Spurred by the ideas of engineer Captain (later General) Bruce Clarke, Engineer Board studies perfected a new steel treadway bridge constructed on pneumatic floats that would carry heavy modern tanks across the rivers of Europe. By 1943, the Engineer Board produced a tank dozer capable of knocking over substantial barriers while conducting an armored assault.

*Engineers lay pierced-steel plank to construct a runway rapidly in New Guinea, February 1944*

When the Japanese bombed military bases in Hawaii and the Philippines on the morning of December 7, 1941, engineer units that had already been deployed to those islands were called upon to respond. The 34th Engineers, a combat regiment that had lost some equipment but incurred no casualties during the bombing in Hawaii, worked to maintain roads that were suffering from heavy military traffic. The skimpy, 1,500-man U.S. Army engineer garrison in the Philippines was almost evenly divided between Filipino and American personnel. After Japanese forces landed there on December 10, the engineers destroyed bridges from one end of Luzon to the other to slow the enemy's advance. The engineers later erected a series of defensive lines on the Bataan Peninsula and fought as infantry in these defenses before succumbing to superior Japanese forces in April and May 1942. In the southern Philippines, a number of U.S. Army engineers escaped to the mountains of Mindanao, where they worked with Filipino guerrillas and

*The Pentagon under construction*

remained active throughout the Japanese occupation.

On the home front in December 1941, the U.S. Army Corps of Engineers assumed the military construction role formerly held by the Quartermaster Corps, and accelerated construction of military bases, including all of the airfields for the U.S. Army Air Forces. An engineer officer headed the construction of the largest office building in the world, the War Department's headquarters, known as the Pentagon. The Corps established Engineer Replacement Training centers at Fort Belvoir, Virginia; Fort Leonard Wood, Missouri; and Camp Abbot, Oregon, to meet the high demand for combat engineers. Further, the Corps adopted enhanced security measures at sensitive facilities such as the Washington Aqueduct. The Corps also developed, built, and oversaw the implementation of significant logistical systems for war support, such as the movement of petroleum and related products along the nation's waterways. Of note, at the outset of the war, the U.S. Army Map Service was formed under the command of the Chief of Engineers. Among the Corps projects contributing to the war effort was the Bonneville Dam, which supplied the power that eventually generated 25 percent of the Nation's finished aluminum used for aircraft and in other armaments.

U.S. Army engineers first entered combat against German and

(above) Lacking Bailey bridging equipment, the 10th Engineer Combat Battalion "hung a bridge in the sky" using captured timbers to cross this gap in the road at Cape Calava, Sicily, August 1943.

(below) Two 5th Army engineer units building a ponton bridge across the Po River north of Bologna, Italy, April 1945.

Italian forces in North Africa, when American forces landed in November 1942. During the first five months of 1943, a few units of American engineers assisted U.S. Army movements in the broad deserts and fields of Tunisia, clearing enemy mines and building roads from scratch. Prior to the American attacks on Gafsa and Maknassy in the barren plains of southern Tunisia, the 1st Engineer Combat Battalion and a company of the 19th Engineer Combat Regiment built combat approach roads through a no-man's land between the combatants, where the engineers were vulnerable to surprise attacks.

After the Allied victory in North Africa, American and British forces landed first in Sicily and then in mainland Italy during the summer of 1943. Defended by well-equipped and determined German forces, Italy's mountainous terrain and rapidly flowing rivers challenged the road- and bridge-building skills of the Army engineers. The combat engineers particularly distinguished themselves in the fighting at and just south of the Rapido River in the Allied drive north from Naples.

The 48th and 235th Engineer Combat Battalions, assigned to an armored task force under Brigadier General Frank Allen that was ordered to capture Mount Porchia just south of the Rapido, not only removed obstacles and opened sup-

Sgt. Joe Specker

ply lines but also fought as infantry on the flanks of the task force's advance. After enemy fire had substantially reduced the armored infantry units leading this attack, the 48th was ordered to secure the top and sides of the mountain. It was in this effort that engineer Sergeant Joe Specker of Odessa, Missouri, having observed an enemy machine-gun nest and several well-placed snipers blocking his company's progress, advanced alone with a machine gun up the rocky slope. Although mortally wounded by intense enemy fire, Sergeant Specker nevertheless set up and fired his weapon so effectively that the enemy machine gun was silenced, and the snipers were forced to withdraw. With this assistance, the battalion was able to clear the summit of Mount Porchia. Sergeant Specker was honored by a posthumous award of the Medal of Honor.

More than a dozen U.S. Army engineer combat battalions landed on the beaches of Normandy during the Allies' assault landing on June 6, 1944. The engineers cleared the beach obstacles and minefields that the Germans had implanted there and absorbed substantial casualties on Omaha Beach, including the loss of two battalion commanders. Bulldozer drivers, often working in the face of heavy enemy fire, opened exits up narrow draws through the cliffs lining the beaches. Some of the engineers quickly engaged in combat with the Germans alongside assault infantry teams. In one such action, Lieutenant Robert Ross of the 37th Engineer Combat

American engineers lay out roads on a French beach, 1944

Battalion took charge of an infantry company that had lost its leaders and led it and his own engineer platoon up the slopes adjoining Omaha Beach, where they killed forty Germans and captured two machine-gun emplacements.

Engineers clear Saint Lo for traffic from Omaha Beachhead.

Crossing the Seine on a ponton bridge, August 1944

The U.S. Army engineers again provided critical support to the achievement and exploitation of the breakthrough that American forces created in late July 1944 in enemy defenses southwest of St. Lo, France. U.S. Army and divisional engineer troops repaired roads and cleared enemy minefields in and beyond St. Lo with exceptional speed, and they rapidly bridged the small rivers in the area to maintain the Americans' momentum. After the German line had been effectively pierced, armored division engineers constructed the treadway bridges needed by Patton's tanks in the Third Army's quick pursuit of the retreating Germans across northern France. Engineer general service regiments behind them rapidly reconstructed or replaced railroad bridges that had been destroyed by the retreating Germans. In Lorraine, the 130th Engineer General Service Regiment built, under heavy artillery fire, a 190-foot-long double-triple Bailey bridge that Third Army troops used to cross the Moselle at Thionville, France. This bridge had to reach ten feet beyond the specified maximum span of such a bridge, yet it successfully carried heavy American tanks.

The massive German offensive in the Ardennes Forest that began on December 16, 1944, exacted a heavy toll among the sparse American forces surprised in the area. A disproportionate number of those troops were engineers who had been operating sawmills or repairing forest roads, and of necessity, these engineer troops were called upon to fight as infantry. The 81st Engineer Combat Battalion, which had been engaged in road maintenance around Auw, Germany, quickly found itself caught in the center of the powerful enemy assault; within a week, the Germans had captured or killed a majority of its troops despite their determined combat, notably in the defense of St. Vith, Belgium.

Colonel H.W. Anderson's 1111th Engineer Combat Group was headquartered at Trois Ponts, Belgium, right in the path of Joachim Peiper's fast-moving German assault tanks. Despite their inferior numbers, Colonel Anderson's engineers put up a stout and effective resistance that crippled Peiper's force. A minefield was hastily laid by a squad of the 291st Engineer Combat Battalion before Stavelot delayed Peiper's entry into that town overnight. On the following day, December 18, engineers from that battalion helped deflect the German tank column away from the critical petroleum depot near Francorchamps, located on the road to Spa, where the First Army had its headquarters. A company of the 51st Engineer Combat Battalion then diverted the column again at Trois Ponts by blowing the bridges there and defending the village alone until airborne troops could reinforce it. Peiper's tanks eventually ran out of fuel well short of his Meuse River objective, and Peiper's men had to abandon them.

*Assembling a treadway bridge in Belgium, 1945*

*Army engineers sanding a road, Luxembourg, 1945*

Gouldin railroad bridge on the Rhine built by Army engineers in ten days, April 1945

To the south, elements of the 44th, 103rd, and 159th Engineer Combat Battalions delayed portions of the German Fifth and Seventh Armies at the villages of Wiltz, Hosingen, and Scheidgen in Luxembourg, before German forces overwhelmed American positions. Although ultimately unsuccessful, the defense undertaken by these engineer units delayed enemy forces long enough to permit American infantry, airborne, and armored units to come to the defense of critically located Bastogne.

Placing explosive charges to demolish concrete tank barriers on the Siegfried Line, October 1944

Engineer troops also fought before Bastogne, some using antitank weapons with which they had no experience. Private Bernard Michin of the 158th Engineer Combat Battalion waited until an enemy tank came within ten yards of him before having sufficient assurance of his target to fire a bazooka at it. The resulting explosion temporarily blinded him. He rolled into a ditch and, hearing enemy machine-gun fire, lobbed a hand grenade toward its source. The firing stopped abruptly. Private Michin was awarded a Distinguished Service Cross.

In January 1945, American forces pushed a badly weakened German army out of the Ardennes and advanced to the river barriers of the Roer and Rhine. Relying on U.S. Army engineer bridging skills, the Americans crossed the Roer on February 23, 1945, before flood waters released by the breaking of upstream dams had subsided, thus surprising the Germans and permitting a rapid American advance.

Engineers also played a critical role in the unexpected capture of the Ludendorff Railroad Bridge across the Rhine at Remagen on March 7, 1945. As elements of the armored combat command, under career engineer officer Brigadier General William M. Hoge, Jr., approached the bridge that afternoon, the Germans set off a charge of dynamite in an unsuccessful attempt to destroy the span. Risking a new explosion, Lieutenant Hugh Mott, Sergeant Eugene Dorland, and Sergeant John Reynolds, all members of Company B, 9th Armored Engineer Battalion, ran onto the bridge in the company of assault infantrymen. The engineers first located four thirty-pound packages of explosives tied to I-beams under the decking, cut them free, and sent them splashing into the Rhine. After the infantry had cleared the far-shore bridge towers, Sergeant Dorland found the master switch for some five hundred pounds of intended bridge demolition explosives, and he quickly shot out the heavy wires leading from it. Under continuing heavy enemy fire, Lieutenant Mott then directed the repair of the bridge's planking, and seven hours later, he reported that tanks could cross.

While nine U.S. Army divisions crossed the Rhine at Remagen, most

*Engineers assembling a Bailey bridge to put across the Rhine River at Wesel, March 1945*

*Roosevelt Bridge over the Rhine*

U.S. forces crossed that broad river in assaults in late March 1945 that were supported by the combat bridge-building endeavors of the U.S. Army Corps of Engineers. Engineer boatmen piloted Navy landing craft to carry assault units across the swift-flowing Rhine. Behind them, other engineers began installing numerous heavy ponton and treadway bridges that would securely tie the assaulting troops to their sources of supply. Third Army engineers built a 1,896-foot-long treadway bridge across the Rhine at Mainz under combat conditions. Further south, Seventh Army engineers completed, in less than ten hours, a 1,047-foot ponton bridge across the Rhine at Worms.

Heavy enemy fire delayed completion of some bridges and exacted casualties. Captain Harold Love, commander of an engineer treadway bridge company, was killed when the treadway section he was ferrying to a partially completed bridge at Milchplatz was struck by a German shell. Nevertheless, the U.S. Army engineer efforts achieved remarkable results. After crossing the Rhine, the Western Allies pushed rapidly across Germany toward their rendezvous with the Russians at the Elbe River. When the Soviet Red Army arrived in Magdeburg in May, it found that Ninth Army engineers had already, on April 13, 1945, built a treadway bridge across the Elbe at Barby fifteen miles south of that eastern German city.

In the fighting against Japanese forces in the Pacific, U.S. Army engineers distinguished themselves

*An engineer soldier of the 96th Engineer Battalion, an African-American unit, operating a bulldozer to construct a reservoir near Port Moresby, New Guinea, February 1943.*

notably during the amphibious landings that they supported. The engineer boat and shore regiments of the 2d, 3rd, and 4th Engineer Special Brigades directed a series of landings on the north coast of New Guinea and on nearby New Britain, Los Negros, Biak, and Morotai Islands as U.S. and Australian forces advanced by sea in a step-by-step fashion toward their October 1944 return to Leyte Island in the Philippines. The engineer boatmen, who brought ashore a task force of the 41st Infantry Division at Nassau Bay, New Guinea, on June 30, 1943, found themselves engaged in hand-to-hand combat with a much larger Japanese force assaulting the beaches just one day after the landing. Demonstrating their skill with knife and bayonet, the engineers held their portion of the beach perimeter.

After the Allies captured the Japanese base at Finschhafen three months later, U.S. Army shore engineers operating the beach depot two miles north of that New Guinea town were surprised by a Japanese landing attempt before dawn on October 17, 1943. Here, engineer gunner Junior Van Noy, a nineteen-year-old private from Idaho, refused to heed calls to withdraw from his shoreside machine-gun position, despite heavy enemy attacks on it with grenades, flame throwers, and rifle fire. Van Noy managed to expend his entire stock of ammunition on the fast-approaching Japanese before succumbing to enemy fire. He alone is thought to have killed at least half of the thirty-nine enemy

*Pvt. Junior Van Noy*

*Engineer aviation battalions used heavy equipment such as bulldozers and carryalls to construct airfields for heavy bombers, Kiriwina Island, July 1943.*

troops who had disembarked. Private Van Noy was honored with a posthumous award of the Medal of Honor.

Engineer combat forces also participated in maneuver warfare on land against the Japanese. On May 29–30, 1943, the Japanese, who had been surrounded by U.S. Army forces on Attu Island in the Aleutians, attempted to break through the portion of the American lines held by an engineer combat company, but the Japanese were decisively repulsed. The unit killed fifty-three of the enemy while suffering only one officer killed and one enlisted man wounded in the battle. In the Philippines, the 302d Engineer Combat Battalion, responsible for road maintenance across rice paddies and swamps near Ormoc on Leyte, built or reinforced fifty-two bridges for tank traffic in mid-December 1944, generally working under small-arms and mortar fire, and contributed men and armored bulldozers to flush enemy troops out of their foxholes in the bamboo thicket. In northern Luzon and on Mindanao in the Philippines in early 1945, divisional engineer battalions completed essential road- and bridge-building projects in difficult mountainous terrain that sometimes rose higher than four thousand feet above sea level. The 106th Engineer Combat Battalion on Mindanao constructed

*Laying pierced-steel plank on an airstrip at Nadzab, New Guinea, February 1944.*

a 425-foot infantry support bridge across the Pulangi River; encountering a gorge 120 feet across and 35 feet deep, they blasted out its sides to quickly create a crude rock bridge. Much of the engineer construction work on Luzon and Mindanao was interrupted by enemy fire. Engineer officers also played principal roles in planning for the invasion of Japan.

During World War II, the U.S. Army Corps of Engineers contributed essential military services wherever the Army was deployed throughout the world.

*Unloading cargo in New Guinea*

*Engineers of the 856th Engineer Aviation Battalion, an African-American unit, grading an airfield on Kiriwina Island, east of New Guinea, October 1943.*

*Engineers searching for Japanese mines*

*Building a Bailey bridge, the Philippines, 1945*

# Exploiting Enemy Mistakes: Army Engineers, Meter Beams, and the Advance into Germany

When the Germans withdrew from northern France in the summer and autumn of 1944, they left Cherbourg Harbor a shambles. A massive reconstruction job faced engineers with the American forces who occupied the city. The difficulty of obtaining adequate construction materials from the United States only exacerbated the problem. The situation demanded prompt and ingenious improvisation, and the Advance Section (ADSEC) engineers of the Communications Zone were up to the task.

The enemy had made a big mistake at Cherbourg, and the engineers turned it to their advantage. Colonel Emerson C. Itschner, ADSEC engineer, recalled the situation: "The Germans were kind enough to leave us a lot of very heavy steel beams, one meter in depth and up to seventy-five feet long. We had enough of these to bridge from the piles that we drove back to the seawall."

Exploitation of the mistake did not stop with the reopening of the Port of Cherbourg. The ADSEC engineers noted that all of the beams bore the name of a single steel mill, Hadir, in Differdange, Luxembourg. Right then, Colonel Itschner decided they would head for Differdange. As soon as the town fell, the ADSEC men were there. They were not disappointed: the Hadir plant was intact, and the citizens were eager to reopen it.

After a little repair and cannibalization, Hadir began once again to produce meter beams. In a short time, these beams were put to many important uses, including the construction of massive railroad bridges across the Rhine. Thus did engineer alertness and ingenuity solve a major supply problem.

*Railroad bridge over the Rhine built by Army engineers, April 1945*

*Completed gaseous diffusion plant at Oak Ridge, Tenn., part of the massive construction program managed by the Manhattan Engineer District*

# The Manhattan Project

The Manhattan Project was the United States' effort to develop an atomic weapon during World War II. In three short years, the project brought atomic weaponry from scientific hypothesis to reality. The U.S. Army Corps of Engineers played a major role in the development of the largest single government program undertaken to that date.

Following the discovery of nuclear fission in Germany in 1930, physicists the world over began experimenting to determine if neutrons were released during fission, and if so, how they might be utilized to create a chain reaction. If controlled in a reactor, such a chain reaction would be a great power source. If uncontrolled, it could produce an explosion far greater than any from chemical explosives.

The initial effort to hasten the progress of atomic research in the United States came from the scientific community. A small group of European scientists had settled in the United States after fleeing from Nazism in the late 1930s, and they were well aware of the atomic research being done in Germany. Fearing that Germany would produce an atomic bomb first, they prevailed upon the renowned physicist Albert Einstein to persuade President Franklin Roosevelt to increase funding for atomic research and development.

After America's entry into the war in December 1941, researchers from the Allied nations joined the effort. The Allies drew up formal agreements on atomic cooperation, and established a scientific military intelligence unit to follow German progress in atomic research.

By spring 1942, Allied research had progressed to the point that an atomic weapon actually seemed possible. The National Defense Research Committee, then coordinating atomic research and headed by Vannevar Bush, began to formulate plans for the construction of production facilities. The U.S. Army Corps of Engineers, designated by the committee to oversee the program, provided the technical expertise required for this mammoth construction project.

On June 18, 1942, Major General W. D. Styer, chief of staff for

Army Services of Supply, directed Colonel James C. Marshall of the U.S. Army Corps of Engineers to form a new engineer district. The district was to carry out the Corps' new responsibility for construction for the project. The new district's offices were initially located in Manhattan at the headquarters of the Corps' New York District. The name "Manhattan" stuck. It seemed to be a name that would arouse the least suspicion for the district, the project, and its super-secret mission.

By September, Brigadier General Leslie R. Groves, formerly deputy chief of the Construction Division in the U.S. Army Corps of Engineers, had been named by Secretary of War Henry L. Stimson to direct the entire project. Under Brigadier General Groves's command, the Manhattan Engineer District began a construction effort that would include production sites across the United States and a workforce of 125,000. Major construction projects included the electromagnetic, gaseous diffusion, and liquid thermal diffusion plants at the Clinton Engineer Works in Oak Ridge, Tennessee; the plutonium production plant at Hanford, Washington; the weapons design and production facilities at Los Alamos, New Mexico; and the numerous facilities such as housing, shopping centers, and hospitals to support the large workforce at these remote and undeveloped locations. Scientific direction remained with the National Defense Research Committee within the Office of Scientific Research and Development, headed by Vannevar Bush.

S-50 thermal diffusion plant under construction at Oak Ridge
National Archives

As research continued in autumn 1942, Groves and Marshall began to select sites for the atomic material production plants. The sites all had to be isolated so they could be sealed off for tight security. They all needed great quantities of both water and electricity. An additional site also had to be found where scientists could finally assemble and test the weapons.

On the recommendation of Groves and Marshall, the government purchased 83,000 acres of land near Clinton, Tennessee, for the Clinton

Oak Ridge, Tenn., shopping center (foreground) and district headquarters (background)

Engineer Works (later called Oak Ridge). Here the U.S. Army Corps of Engineers built uranium separation plants to separate the fissionable isotope uranium-235 from the more prevalent isotope in uranium ore, uranium-238. Army engineers also constructed residential communities to house employees.

In December 1942, when famed scientist Enrico Fermi produced a controlled chain reaction at the University of Chicago, he discovered a new material suitable for fission. He found that during the chain reaction, uranium-238 could capture neutrons and be transformed into plutonium, a new element as unstable as uranium-235. Twelve days after Fermi's successful experiment, Groves initiated discussions involving

Plant under construction at Oak Ridge, Tenn., April 22, 1944

*First reactor pile area at Hanford Engineer Works*

leading scientists and industry and Corps representatives to build a plutonium plant site. The government soon purchased almost a half million acres of land around Hanford, Washington, near Bonneville Dam, for the construction of five plutonium reactors and employee housing.

In addition to building huge industrial plants and providing the most basic community needs of water, roads, sanitation, housing, and power, the U.S. Army Corps of Engineers also managed the construction of scientific equipment, newly designed and as yet untried. The initial budget outlay for the atomic energy project in June 1942 was only $85 million. Project requirements had been underestimated. For example, at Oak Ridge the cost of the land alone was $4 million. By the end of 1946, construction costs at Oak Ridge totaled $304 million. Research at this site eventually totaled $20 million, engineering $6 million, and operations $204 million. Power for operations cost $10 million. Instead of requiring a workforce of 2,500 people, as originally estimated, Oak Ridge eventually had 24,000 employees on the payroll.

As work continued at Oak Ridge and Hanford, General Groves appointed J. Robert Oppenheimer to take charge of the newly created weapons laboratory in an isolated desert area around Los Alamos, New

*Maj. Gen. Leslie R. Groves considers potential targets.*

Mexico. Here scientists assembled the actual weapons. The first explosion of an atomic bomb occurred at the Trinity Site in the predawn hours of July 16, 1945. The atomic bomb was a reality, and those meant for actual use were already in transit to the Pacific.

The engineering problems encountered in the project were numerous. Groves and his staff fought constantly for needed raw materials. The engineers had to translate the scientists' theories into precise specifications. New materials had to be formulated for building the reactors and the separation equipment. Contractors were held to extremely exacting specifications for everything they supplied. The Corps' engineering role required the coordination of construction with research and new discoveries. It required building huge industrial facilities along with the housing, community, and recreational facilities needed to provide a livable environment for the employees. It required the transportation of goods to these isolated areas, the management of huge amounts of money, and the coordination of input from hundreds of contractors. Further complicating the development process was the need for secrecy—only a select few knew that the ultimate goal of the Manhattan Project was to produce an atomic bomb.

The project also required the maintenance of a delicate relationship

*Administrative and residential areas at the Hanford Engineer Works*

*Completed chemical separation plants and steam-electric facility at Hanford, Wash.*

between the military and the scientific communities. Workers and scientists had relocated to physically isolated areas and, because of the secrecy of their work, had to limit their contact with the outside world. Even in wartime, when the work had a special urgency and sacrifices were made for the war effort, morale was a great concern. The scientists especially were uncomfortable under military supervision and security restrictions. Very few of the thousands of employees on the project knew what they were actually working on because of the strict security; however, the employees did share anxiety over the unknown dangers inherent in the materials they dealt with.

No one dreamed at the beginning how massive the project would become. The four-year-long research and development project was completed at a cost of $2 billion. Very few who worked on the project understood at the time the tremendous impact the project would have on the world. In the end, the Manhattan Project produced the weapons that leveled Hiroshima and Nagasaki, ending World War II and marking the onset of the Atomic Age.

# Women Played Key Roles in the Manhattan Project

While significant numbers of civilian women served at all of the project sites for the development of the atomic bomb, many of the women serving in the Manhattan Engineer District were Soldiers and officers of the U.S. Army. During World War II, more than 150,000 American women served in the Women's Army Corps, or WAC, and many assigned to the U.S. Army Corps of Engineers participated in the Manhattan Project. As early as 1943, women Soldiers were brought into the Manhattan Project for clerical, technical, and administrative work.

The need for additional personnel led to the establishment of a Manhattan District Women's Army Corps Detachment on June 3, 1943. After February 1, 1945, the entire military complement of the Manhattan District was designated by the Chief of Engineers as the 9812th Technical Services Unit–Civil Engineers. By the end of the war in 1945, approximately 425 women were in this unit, which earned the Meritorious Service Unit Award.

Jobs performed by women assigned to the Manhattan Engineer District included stenographer, telephone operator, laboratory technician, clerk, cryptographer, classified information handler, metallurgist, electronics technician, photographer, spectroscopist, nurse, and scientist. A large number of notable women, both WAC and civilian, worked in the Manhattan Project. The first commanding officer of the WAC detachment was Lieutenant Frances W. House. She was succeeded by Lieutenant Arlene G. Scheidenhelm in March 1944. Master Sergeant Elizabeth Wilson ran the cyclotron at Los Alamos. Electronics technician Jane Heydorn helped to develop bomb-testing equipment. Lieutenant Catherine Piccolo wrote official press releases explaining why the bombs were utilized. Physicist Chien Shiung Wu played a key role in developing the gaseous diffusion uranium separation process. Leona Woods monitored the first nuclear chain reaction. The head of a vital research team, Maria Goeppert Mayer, later received the Nobel Prize in physics. Elizabeth Riddle Graves developed a neutron reflector to surround the atom core at Los Alamos.

In commending the WACs for their contributions to the Manhattan Project, on August 9, 1945, then-Major General Groves wrote, "I wish to express to you, the military personnel of the Manhattan Project, my official and personal appreciation for the industry, ability and attention to duty under most trying conditions which you have displayed since the inception of the project. Without you, this project could not have achieved success. Your devotion to duty and particularly your conscientious efforts to maintain the vital security of the project have been of the highest order. You have every right to be proud of the vital role which you have played in this development which has culminated in the use in combat against Japan of the greatest weapon man has ever forged. Our achievement could not have been realized but for your individual effort. The saving in American lives will be your reward."

*Women's Army Corps Detachment at Oak Ridge, Tenn.*

*Soldiers of the 2d Engineer Combat Battalion sweep a road for anti-tank mines, March 1953*
National Archives

# Army Engineers in Korea

Following World War II, the Korean Peninsula was occupied by the victorious Allies. By the time the occupation ended, two Korean governments had arisen—the Soviet-sponsored Democratic People's Republic of Korea in the north and the Western-supported Republic of Korea in the south. On June 25, 1950, the North Korean government launched an attack across the 38th parallel in a plan to unite the peninsula under communist rule.

Surprised by the North Korean attack, U.S. Army troops in Korea and the Republic of Korea's forces could at first do no more than delay the advance of the larger and better equipped North Korean forces. U.S. Army engineers played a major role in this delaying action, mining roads and destroying key bridges. The rugged terrain of the Korean Peninsula and the numerical superiority of enemy forces made engineer construction and combat vital to the U.S. Army during the Korean War.

In the early fighting, engineers were frequently required to do tasks not traditionally theirs. For example, on July 20, 1950, members of Company C, 3rd Engineer Combat Battalion, made the first verifiable combat use of the newly developed 3.5-inch rocket launcher, using it to destroy a tank that was threatening their division commander near Taejon. Attempting to withdraw from Taejon that evening, U.S. forces were stopped for a time by enemy roadblocks. Engineer Sergeant George Libby placed wounded men on an artillery tractor and used his body to shield the driver as it crashed through two enemy roadblocks before

*Engineers mine a bridge to impede the North Korean advance, July 1950*
*U.S. Army Engineer School*

Sgt. George Libby

reaching American lines to the south. Sergeant Libby, who died of his wounds, was posthumously awarded the Medal of Honor.

By early August 1950, U.S. and South Korean forces had withdrawn to the southeastern port city of Pusan. The outnumbered allied forces maintained a long defensive perimeter around Pusan as General Douglas MacArthur prepared to land a large body of U.S. troops behind enemy lines at Inch'on. Engineers were frequently committed to fight as infantry on the Pusan perimeter. Private Melvin Brown of the 8th Engineer Combat Battalion was awarded the Medal of Honor for bravely holding his position on a wall of the ancient fortress of Kasan during an enemy assault. After he had expended his ammunition, Private Brown used his entrenching tool to repel the armed attackers as they reached the top of the wall.

MacArthur's behind-the-lines assault at Inch'on, which began on September 15, 1950, caught the enemy by surprise. Subsequently, U.S. forces took the offensive throughout Korea. The bridge-building and road and rail repairs undertaken by the U.S. Army engineers allowed U.S. and allied forces to push north rapidly in pursuit of the disintegrating North Korean

Soldiers of the 14th Engineer Battalion place barbed wire along the banks of the Naktong River, 10 August 1950.
U.S. Army Engineer School

Engineers of the 2d Infantry Division construct a bypass to enable heavy equipment to cross the Hwang-gang River, 25 September 1950

U.S. Army Engineer School

Army. Handicapped at first by tremendous shortages of supplies, the engineers had to make innovative use of available materials for these construction efforts.

When Chinese units began their powerful counteroffensive in November 1950, the U.S. Army engineers had to destroy many of the same bridges they had recently built as U.S. forces again retreated south of Seoul. But lateral roads built by the engineers behind the new defensive lines proved critical when the Chinese broke through a portion of that line. These roads enabled the

Men of the 65th Combat Engineer Battalion reinforce a muddy road on the north bank of the Han River, March 1951

U.S. Army Engineer School

*A cable car built by the 3d Engineers carries men and supplies up the steep hillsides*
*National Archives*

*Soldiers of the 77th Engineer Combat Company lay a single-apron barbed wire fence*
*National Archives*

Americans to transport the 3rd Infantry Division 100 miles in a single day to plug the hole that the Chinese had created.

As U.S. forces returned to the offensive in mountainous central Korea in early 1951, engineer units blasted cliffsides to construct new roads and built aerial tramways to carry supplies to the troops. When the advancing 23rd Regimental Combat Team and a French battalion were surrounded at Chipyong-ni on February 13, 1951, by an attacking force apparently composed of three Chinese divisions, the engineer company supporting the combat team fought as infantry. They withstood the attack until an American armored relief column could reach the town two days later.

In early October 1951, the 2d Engineer Combat Battalion converted a rough track leading north to Mundung-ni into a road usable by armor, enabling an American tank battalion to surprise a Chinese column attempting to relieve hard-pressed Chinese troops on Heartbreak Ridge near the 38th parallel. A U.S. Army engineer construction battalion also supported the 1st Marine Division in its combat in mountainous central Korea during much of 1951.

The engineers confronted a critical challenge after the summer floods of July 1952 washed out two

*Soldiers of the 185th Engineer Combat Battalion stand watch over a floating bridge damaged by flood waters on the Soyang River, May 1951.*

*National Archives*

*Built by the 84th Engineer Construction Battalion, the Libby Bridge provided a vital high-level crossing of the Imjin River, July 1953.*

*National Archives*

of the five high-level bridges across the Imjin River, located a mere four miles behind the battle lines of three U.S. Army divisions. After installing two temporary floating bridges, engineer troops built at the less critical site an innovative low-level bridge sturdy enough to survive if overtopped by flood waters. In the center of the I Corps line, within range of enemy artillery, the 84th Engineer Construction Battalion erected a modern, commercial-type highway bridge utilizing sheet-pile cofferdams and reinforced concrete piers. Dedicated to engineer Medal of Honor recipient Sergeant George Libby, that bridge remains in use and retains its tactical significance decades after its construction.

The U.S. Army engineers in Korea compiled a remarkable record of combat and wartime construction that complemented and often multiplied the combat effectiveness of the highly motorized and mobile American units engaged there. U.S. Army engineers often were the unsung heroes of the Korean War, for they helped create the environment that allowed the United States and its allies to fight and win.

*On their way to a second tour of duty in Korea, soldiers of the 8th Engineer Combat Battalion wait to disembark from their troop transport.*
*National Archives*

# In Their Own Words: The U.S. Army Engineer Experience in Korea

The Korean Peninsula was an inhospitable place in which to wage a war, not only due to topography and climate but also because the U.S. Army faced a well-supplied enemy fighting an ideological crusade. In overcoming the elements as well as a tenacious enemy, U.S. Army engineers again proved invaluable in combat support roles. Personal accounts by some of the participants shed light on the challenges they faced.

Engineers were deeply involved with operations in Korea before the outbreak of hostilities. After reading intelligence reports, Lieutenant Colonel Edward Rowny, a planner in General Douglas MacArthur's Far East Command (FECOM) headquarters, warned intelligence officials that the United States needed to be mindful of the possibility of an attack in Korea. After the North Koreans invaded, and U.S. and South Korean forces withdrew south, Rowny and others in FECOM helped draft a plan for an amphibious invasion to relieve the pressure on the Pusan perimeter. The staff officers recommended invading near or slightly behind the front line. MacArthur took a much more aggressive approach, directing his staff to study an invasion at the port of Inch'on, 100 kilometers up the coast opposite Seoul. "One should land as close as possible to the objective, and the objective is the capital" the General said. "You're all timid," MacArthur lectured his staff, "you should think boldly and decisively." When another planner cited the danger posed by Inch'on's thirty-one-foot tide, MacArthur brushed those fears aside. "And as for the tides," he said, "don't take counsel of your fears. Physical obstacles can be overcome by good planning, strong nerves and will power." Rowny would need all those attributes, for General MacArthur appointed the young officer

*Lt. Col. Edward Rowny is awarded the Legion of Merit by Maj. Gen. Edward Almond, commander, X Corps, December 1950.*

National Archives

*Snow and bitter cold made operations in Korea difficult. Here, soldiers of the 2d Engineer Combat Battalion survey a new supply route.*

National Archives

to be the engineer for the Inch'on landing and he went ashore in the first wave of the assault.

During the first winter of the Korean War, Lieutenant Maurice D. Roush was a platoon leader with the 13th Engineer Combat Battalion. He described the lack of personal equipment to face the harsh seasonal conditions following his amphibious debarkation along the eastern coast of Korea: "About the time we landed we were given trigger-finger mittens and some hats with earflaps. That was the extent of winter gear. We still had our blanket sleeping bags. We didn't have good parkas or good footgear. We got into one of the worst winter situations I've ever seen. I've never been so cold—and I come from Wyoming! Up in North Korea on the plateau, up near the Yalu River, it's extremely cold."

*Republic of Korea Army Engineer Training School*

*National Archives*

For most of 1952, Lieutenant Colonel Harry D. Hoskins, Jr., commanded the 10th Engineer Combat Battalion in support of the 3d Infantry Division near the 38th parallel. He later recounted the defensive measures Army engineers used: "We made a series of firetraps to be used in the event the North Koreans got into the Ch'orwon Valley. That was a wide area, so we needed to have a lot of people or a lot of mines or something to stop them. You have to have a series of interlocking firetraps to stop that kind of an attack. At that time the North Koreans didn't have tanks. They were just waves, and waves, and waves of manpower. You had to have mines, especially antipersonnel mines, to stop the manpower and any heavy vehicles. Then all kinds of napalm were needed, so you could drop it in quickly. You couldn't be waiting around because once there was a breakthrough they'd pour in there in a hell of a hurry."

Colonel Pashal N. Strong, Jr., was an engineer officer with the Eighth Army. Commenting on the performance of reserve engineer officers, he noted, "From my own experience, the best regimental commanders for heavy construction work were contractors who had been doing that in the reserves. I found them better for that than the West Point graduates, because the West Point graduates hadn't had the practical experience in heavy construction that the contrac-

*An engineer uses a bulldozer to repair a road damaged by retreating enemy troops.*
*National Archives*

tors had. West Pointers also were a bit too worried about the spit-and-polish, sometimes at the expense of their construction activities."

Personnel shortages forced the U.S. Army to use Korean soldiers to fill out many of its under strength units. The Korean soldiers were introduced into the U.S. units through the Korean Augmentation to the United States Army (KATUSA) program, and the Korean soldiers quickly proved their value. Although the KATUSAs had to be brought up to speed, once trained they proved invaluable to the U.S. Army engineers. As Lieutenant Colonel Evan S. Pickett later commented, "When we first received them, the KATUSA troops were untrained and inadequate for engineer work. They had no coordination for running bulldozers and graders or running our hydraulic equipment. They were good at hand labor, but they were very poor with mechanical equipment. But, as time went on, we found that they learned to operate the mechanical equipment fairly quickly.… In the end they were well qualified and seemed to contribute a lot to our mission."

Lieutenant Joseph K. Bratton served with the 13th Engineers, 7th Division. Lieutenant Bratton, who later became Chief of Engineers, summed up the importance of his experience in Korea this way: "The overwhelming positive lesson I learned was the great value of direct engineer support to the infantry regiments. If the regiment knew how to use the engineers, and if the engineers were not too bashful in explaining their capabilities to the tactical unit commanders, they gained a great deal from the engineers' support. I was thrilled to see how well our companies worked with the regiments. That was happening when I arrived and it built up while the 7th Division stayed in Korea. That was a tremendous lesson that I think not only engineers learned, but everybody learned."

*Soldiers of the 299th Engineer Battalion check the alignment of piles before they are driven, May 1966.*

# U.S. Army Engineers in the Vietnam Conflict

The U.S. Army again called upon its engineers for combat support in Asia to assist the Republic of Vietnam in its struggle against a communist insurgency. Beginning in the early 1960s, the American commitment of ground forces to Vietnam eventually numbered more than 535,000 and lasted for a decade. In South Vietnam, insurgent forces often relied heavily upon a strategy of concealment when in combat against American troops. U.S. Army operations in Vietnam thus did not occur along a well-defined front line, but could break out wherever the Americans encountered Viet Cong guerrillas or North Vietnamese troops. The elusiveness of the enemy led U.S. Army engineers to alter the way they pursued their task of enhancing the combat effectiveness of friendly forces.

American forces frequently employed search-and-destroy missions to attack areas of enemy strength. The 1st Engineer Battalion supported Operation Rolling Stone in Binh Duong Province near Saigon by building a road into the Iron Triangle and War Zone D, two staging areas frequently used by the Viet Cong. Men of this battalion engaged in a half-hour-long firefight with the enemy on February 26, 1966. The following summer, a fifty-two-bulldozer battalion task force cleared 2,700 acres of jungle, destroyed six miles of enemy

*Land clearing at Ben Cat, South Vietnam*

Soldiers of the 1st Engineer Battalion sweep for mines near the village of Thien Thanh.

tunnels, and demolished eleven factories and villages in the Iron Triangle.

The widespread use of helicopter transport in Vietnam enabled U.S. forces to respond quickly to attacks anywhere in the country. After South Vietnamese forces relieved a besieged Special Forces camp at Plei Me in the Central Highlands in October 1965, an engineer company of the airmobile 1st Cavalry Division lengthened and improved an earthen airfield at a nearby tea plantation, using equipment brought in by helicopter. The division then pursued the attacking North Vietnamese regiments west from Plei Me through the jungles of the highlands. For forward supply and reinforcements in this campaign, the division relied upon helicopter landing zones that divisional engineers quickly cleared from the jungle using chain saws and demolitions. By the time the North Vietnamese forces reached the safety of Cambodia, they had lost 1,800 men. During the next ten months, the 8th Engineer Battalion built seven airfields for the division in the Central Highlands, including one at a site eight miles from the Cambodian border to which all construction equipment, supplies, and personnel had to be transported by helicopter. Moving the equipment by air was possible because U.S. Army engineer planners had modified procurement orders for large earthmoving equipment to obtain machinery that could be disassembled for airlift and then quickly reassembled.

Various technological innovations aided the U.S. Army engineers in Vietnam. To combat the thick mud that could quickly disable the tactical airfields in the monsoon season, the engineers employed the new T-17 membrane, a neoprene-coated fabric used to cover the airfields and provide them with an impermeable "raincoat." Another problem was the additional wear on helicopter rotors caused by the abrasive dust stirred up by flight operations. The swirling man-made dust storms also significantly reduced helicopter pilots' vision, further complicating flight operations. At the end of 1965, U.S. Army Chief of Staff General Harold K. Johnson directed Lieutenant General William F. Cassidy, the Chief of Engineers, to find a solution. Cassidy relied upon the expertise of the Corps' research laboratories, which had been using peneprime, a dust palliative with an asphalt base, as a penetrant in civil works projects. Personnel from the Waterways Experiment Station led an assessment team to Vietnam to determine the appropriateness of this agent for battlefield use. Subsequently, U.S. Army engineers sprayed peneprime onto heliport sites during the dry season to prevent dust clouds from interfering with helicopter operations.

*Spreading T-17 membrane on a runway*

Land clearing was a very effective weapon against the Viet Cong insurgency. Guerrilla forces used the thick forests along the nation's major transportation routes to conceal themselves before laying mines or staging ambushes. Consequently, the engineers had to clear all vegetation

*Installing T-17 membrane at Bao Loc*

*Department of Defense*

60th Land Clearing Company's Rome plow

A heavily armed Rome plow operator clearing jungle

up to 100 yards on either side of major roadways. Finding bulldozers and flammable napalm unequal to the task, in 1967 the engineers introduced the Rome plow, a military tractor equipped with a protective cab and a special tree-cutting blade that was sharpened daily. The Rome plow was used to cut trees at or near ground level; it also had a stinger to split longer trees. Lieutenant General Julian Ewell, commander of the 9th Infantry Division in Vietnam, called the Rome plow "the most effective device" in his arsenal. A land-clearing engineer company equipped with thirty Rome plows could clear 180 to 200 acres of medium-density jungle each day.

Supporting the U.S. military effort in Vietnam required a massive construction effort. During the war, U.S. Army engineers, supported by a large contractor workforce, built thousands of facilities including warehouses, piers, troop cantonment areas, maintenance facilities, roads, and bridges. At its peak, Army engineer troop strength in Vietnam approached 40,000 soldiers, augmented by tens of thousands of contractors. The presence of so many construction contractors was a notable innovation and marked the first time civilians assumed a major construction role in an active theater of operations.

When American troops and equipment began to pour into

Vietnam in the mid-1960s, the country had only two ports capable of docking oceangoing vessels. With 90 percent of U.S. supplies destined for Vietnam arriving by ship, the lack of sufficient port facilities soon created a massive backlog of ships waiting to unload. To ease the congestion, the United States began improving South Vietnam's ports. To improve access, a fleet of dredges, including two from the Corps of Engineers, set to work clearing waterways and deepening channels. To expedite the construction of deep-water berthing facilities, Army engineers installed floating piers. Fabricated by the DeLong Corporation in the United States, the first pier and all of its support equipment were towed to Vietnam and installed by the 497th Port Construction Company. The pier consisted of a ninety by three-hundred-foot-long barge supported by eighteen tubular steel caissons to anchor it to the bottom. Once caissons were in place the engineers used pneumatic jacks attached to the caisson collars to lift the barge up on its legs to the right height. Engineers installed the first DeLong pier at Cam Ranh Bay in December 1965, and it quickly doubled the capacity of the port. Soon after, the DeLong piers were installed at many of South Vietnam's major ports.

The enemy's Tet Offensive early in 1968 closed for more than a month several critical roads, particularly in the northern part of the Republic of Vietnam. The U.S. Army's 35th Engineer Battalion, which had concentrated on road-building during its previous service in Vietnam, reopened coastal Route 1 north of Da Nang in late February 1968 while assigned to the

*The port of Cam Ranh Bay showing newly constructed piers.*

*DeLong pier floated into position with caissons ready to be driven down, Cam Ranh Bay, December 1965.*

*Soldiers of the 1st Engineer Battalion driving pile for the construction of a new bridge near Di An*

III Marine Amphibious Force. By this time, the engineers had built a sufficient number of airfields, heliports, and troop cantonments to permit them to continue to concentrate on road construction. The 27th Engineer Battalion built a new, all-weather highway from Hue west to the A Shau Valley, an enemy stronghold.

In fact, U.S. Army engineers constructed much of South Vietnam's highway system. Overall, engineer troops constructed roughly 900 miles of modern, paved highways connecting the major population centers of the Republic of Vietnam. Engineer officers also monitored the construction by private American contractors of an additional 550 miles of Vietnamese highways. Brigadier General Carroll Dunn, Director of Construction, Military Assistance Command, Vietnam, described the road construction effort as "the single most effective and important development program undertaken by the American effort in Vietnam." The engineers also safeguarded the roads. Units in the Mekong Delta developed a clay-lime coagulation process that they used there to build durable roads from locally available materials. The engineers protected their bridges by installing extensive lighting systems and antiswimmer and antimine devices using concertina wire and booms.

Army engineers also undertook certain responsibilities for installation security, and these sometimes involved heroic individual actions. When an enemy team infiltrated the base of the 173d Engineer Company at Camp Radcliff at An Khe in the Central Highlands on March 20, 1969, engineer Corporal Terry Kawamura threw himself on an explosive charge that had been hurled into his quarters, absorbing its blast and thereby protecting other members of his unit endangered in the attack. Corporal Kawamura was posthumously awarded the Medal of Honor.

A half dozen U.S. Army engineer battalions participated in the Cambodian incursion in May and June of 1970. Engineers built thirty-five miles of new roads, twenty-three fixed bridges, and twenty-five fire-support bases during the attack on North Vietnamese supply points and staging areas within Cambodia. During this period, the senior U.S. Army engineer officer in Vietnam, Major General John Dillard, and two other high-ranking engineers were killed when their helicopter was shot down southwest of Pleiku. The U.S. Army Corps of Engineers showed the same bravery and dedication as the combat troops during service in Southeast Asia.

*Engineers pour concrete for a new bridge approach.*

# U.S. Army Engineers Helped Clear Viet Cong Tunnels

To counter the immense technological advantage held by U.S. and allied forces during the Vietnam conflict, the Viet Cong developed an extensive network of underground tunnel complexes. From these tunnels, which were concentrated mostly around Cu Chi but spread as far as the outskirts of Saigon, the enemy could ambush American forces and then safely vanish underground. The tunnels became so highly developed that they eventually contained armories, hospitals, mess halls, manufacturing centers, and storage facilities. Some complexes ranged up to fifty kilometers long. Extensive booby-trapping made it next to impossible for American troops to extricate the enemy from their underground safe havens, which allowed them to withstand intense aerial bombardment.

U.S. Army engineers developed a number of methods for destroying the tunnels or making them unusable. The least effective was by mechanical means, as bulldozers and plows could displace only the shallowest tunnels. Moreover, it was difficult to deploy bulldozers and plows in densely vegetated and remote areas. Flooding also proved substantially ineffective because the Viet Cong had dug additional wells deep inside the tunnel complexes to prevent them from becoming saturated. An even less desirable—but most immediately available—method was for volunteers from special engineer tunnel demolition teams (who became known as "tunnel rats") to enter the tunnels headfirst to clear them out the hard way.

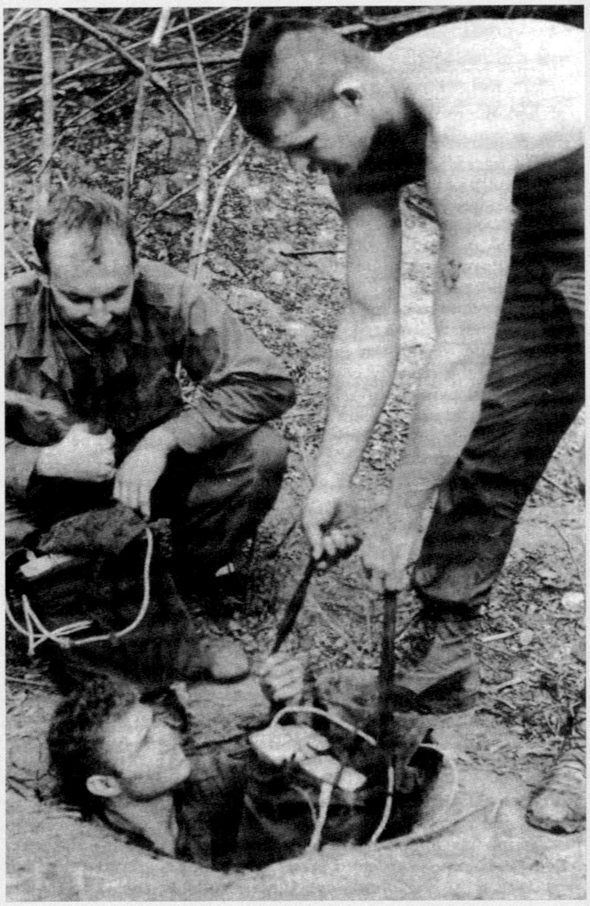

*An engineer tunnel demolition team*
Department of Defense

Conventional explosives also were used to clear the tunnels. Block explosives placed at critical points with a force of two pounds per foot could bring down a section, and shaped charges facing upward could destroy certain tunnel segments. Another method was to deposit cratering charges in five-foot-deep holes along the outside trace of a known tunnel. Because of their explosive characteristics, Bangalore torpedoes were the most successful conventional means of effecting complete destruction, but each section had to be carried into the tunnels and emplaced by hand.

Other methods employed were innovative. One was to run tubing along the length of a tunnel and then fill it with liquid explosive either by gravity fill or a pumping system, although the highly flammable nature of these liquid explosives often countered their effective use. Another means of denying use of the tunnels was through the introduction of tear gas dispersed by the "Mitey-Mite" blower. Although these chemical agents could persist on the walls of the tunnels and render them uninhabitable for months, the dense jungle and attendant climatic conditions often "swallowed up" chemical dispersants.

In the most effective method, engineers used acetylene for destruction of tunnels with less than seven feet of overburden. Three cubic meters of acetylene pumped into an area could destroy forty cubic meters of tunnel volume. When acetylene was used in conjunction with conventional explosives, the effect could collapse fifteen feet of overburden. In the end, however, enemy operations from the tunnels were never completely eradicated.

*Engineers setting charges to collapse underground enemy bunkers*

*A Saturn V test vehicle emerges from the Vehicle Assembly Building. The launch control center is in the foreground.*

# The Corps and the Space Program

Given its past experience in missile site construction on the Intercontinental Ballistic Missile (ICBM) program, the U.S. Army Corps of Engineers was the logical choice of Congress and the National Aeronautics and Space Administration (NASA) to oversee NASA's accelerated construction program in the early 1960s. Not only was the Corps well versed in missile facility construction, using the Corps also eliminated the need for NASA to establish its own construction organization.

Although the Corps had been providing NASA with design and construction services since the spring of 1960, the scope of the Corps' support changed dramatically in May 1961 when President John F. Kennedy declared a national goal of landing a man on the Moon and returning him safely to Earth within the decade. The president's speech was the genesis of the Apollo Program, and the following September the civilian space agency turned to the Corps to build the facilities that would become the hub of the Nation's space program—the sprawling Mississippi Test Facility, later

*Carrying an Apollo spacecraft, a Saturn V launch vehicle takes off from Kennedy Space Center.*

A specially configured barge carries a Saturn booster near the Mississippi test facility.

renamed the John C. Stennis Space Center; the Manned Spacecraft Center in Houston, now the Lyndon B. Johnson Space Center; and the 84,000-acre facility on the east coast of Florida that would later be named the John F. Kennedy Space Center.

In response to the president's mandate, NASA and the Corps embarked on a massive construction program along the Gulf of Mexico and the Atlantic Ocean, an area that quickly came to be called the "NASA crescent." The launch vehicles destined to carry the NASA astronauts into space were orders of magnitude larger than NASA had ever built, and consequently transporting them by water was the only feasible alternative. As a result, early in the site construction process planners decided that it was imperative that all of the new facilities have easy access to navigable waterways to transport the boosters for testing and launch. Indeed, proximity to water was a factor in the selection of Houston as the site for the manned spacecraft center. On September 25, 1961, only three days after NASA requested the Corps' assistance, the Fort Worth District began arranging preliminary topographic and utility surveys of the site of the manned spacecraft center.

Fort Worth District's experience with incremental funding stood NASA in good stead in the construction of the center. This method of funding was based on the congressional tradition of appropriating construction funds on a year-to-year basis. That meant the district contracted for each segment of the center as a separate unit. One virtue of this procedure was that it allowed significant changes in construction plans without delaying the project. For instance, on July 17, 1962, NASA announced that the future Mission Control Center would also be located at the Houston center. This decision forced the Corps to insert an entirely new building into its master plan.

The incremental funding system also permitted major modifications of facilities already under construction. This was important because speed was essential if NASA's goals were to be met, and the engineers and NASA had to construct buildings at the same time NASA was designing the laboratories and machines they would contain. Troubles with the Space Environment Simulation Chamber showed the value of the arrangement. The failure of the chamber during its first vacuum test required not only its redesign, but also numerous changes in the one-third-completed building. Incremental funding enabled contract modifications to be made without major delays. In November 1966, after spending some $75 million on the 1,600-acre project, Fort Worth District completed its work on what came to be called the Johnson Manned Spacecraft Center.

Mobile District's involvement in NASA's rocket testing program began with the transfer of the Army Ballistic Missile Agency's Development Operations Division at the George C. Marshall Space Flight Center at Redstone Arsenal in Huntsville, Alabama, to NASA in 1959. NASA then established the Michoud Assembly Facility near New Orleans as a support facility for the Huntsville projects. Michoud was the assembly plant for the large Saturn booster rockets. In autumn 1961, NASA established its test facility for the rockets assembled at Michoud on a 217-square-mile tract at the Mississippi Test Center, later known as the National Space Technology Laboratories, accessible from Michoud by both land and water. Mobile District spent more than $200 million constructing space program

*The Manned Spacecraft Center, Houston, Texas, under construction, December 1966.*

*Constructed at the George C. Marshall Space Flight Center in Huntsville, Ala., the Saturn V test stand was designed to withstand 7.5 million pounds of thrust.*

A Corps official poses with drawings and specifications from the mammoth project at Launch Complex 39.

facilities up to the completion of the test center in April 1966. The center's initial mission was to test the Apollo-Saturn V second stage booster and to test flight models of both the first and second stage boosters, with thrusts of 7.5 million and 1 million pounds, respectively. The site became NASA's principal test facility.

Initially, design and construction work at Kennedy Space Center was handled by the Jacksonville District, but to meet the demands of the Apollo construction program in May 1963, the Corps of Engineers established the new Canaveral District to handle the construction effort.

Perhaps no other structure better symbolizes the Corps of Engineers' contribution to the United States space program than Launch Complex 39 at the Kennedy Space Center. Built to assemble and launch the giant Saturn V rockets that would carry the Apollo astronauts to the moon, facility construction began in 1963. Major components of the launch complex included the Vehicle Assembly Building (VAB), a 525-foot-tall building where the rockets were assembled; the adjacent launch control center that included four

The Vehicle Assembly Building at the Kennedy Space Center. Components for the Saturn V launch vehicle arrived by barge in the basin (foreground).

*Launch Complex 39, Pad A, with the crawlerway connecting the pad to the Vehicle Assembly Building in the distance.*

command centers; and a three-mile-long crawlerway built to transport the Saturn V rockets to the launch pad. The launch complex contained two launch pads, 39A and 39B, and each covered a quarter square mile. But the launch complex was only part of the project; supporting the NASA program was a large contractor work force, and to house them the Corps constructed an industrial area on nearby Merritt Island that encompassed fifty buildings, thirty-eight miles of roads, and at its peak 14,000 employees worked there.

Ultimately, the Kennedy Space Center cost $900 million to build, and in the decades since its completion has served as America's gateway to space. In the words of NASA Administrator James Webb, "The road to the moon is paved with bricks, steel and concrete here on earth."

Other Corps offices completed additional construction for NASA. For example, the New England Division selected the site for and supervised the construction of the Electronics Research Center in Cambridge, Massachusetts, in the late 1960s. That facility is now the Volpe National Transportation Systems Center. In supervising more than $1 billion of NASA construction, elements of the U.S. Army Corps of Engineers in all parts of the country made major contributions to the national space effort.

# Lunar Maps for NASA

As the U.S. Government looked toward manned spaceflight and an eventual trip to the moon, it became clear that astronauts would need concise maps of that terrain. In 1958, the Army Map Service of the U.S. Army Corps of Engineers began to assess the feasibility of producing an accurate map of the moon based upon telescopic photographs. These Corps topographers concluded that reconnaissance-type photomaps at the scale of 1:5,000,000 were feasible; however, such maps would show only the most general of terrain features.

There were considerable technical challenges to this topographical effort. The moon was a quarter million miles from the Earth. Virtually all photomaps of the moon were taken from an altitude of six miles above the Earth. Because all photomaps were nearly identical, there was no way to utilize stereoscopic techniques to form three-dimensional images that could determine elevations for terrain features. Furthermore, there were no established fixed reference points on the moon by which explorers could determine the elevation, latitude, and longitude of their location.

To overcome initial failed attempts, topographers developed new or improved techniques and equipment. An important innovation was the use of closed-circuit television to enable mapmakers to observe lunar features under different conditions of light and shadow. This process made it possible to determine accurately the height and depth of various terrain features.

The resultant lunar map represented the visible surface of the moon at the feasible scale and showed five thousand geographical features. These terrain features were shown with 1,000-meter contours, and in some cases with 500-meter contours. The Corps managed to map certain small areas in greater detail; for instance, proposed NASA landing sites were mapped at a scale of 1:250,000, with color tinting added for realism.

The Army Map Service also produced rubber or plastic three-dimensional models of parts of the moon's surface. These models were photographed and the films made from them, when projected on large screens, effectively portrayed the varying altitudes that astronauts would face. The models were used in simulated landings practiced at NASA experimental stations.

*First moonwalk*

*National Aeronautics and Space Administration*

The topographic engineers also found solutions to other problems plaguing the space program. They developed a material that could withstand the extreme conditions of space travel and exposure on the moon. Special plastic and rubber compounds allowed the development of foldable maps that could withstand temperatures ranging from –250 to 214 degrees Fahrenheit. Additionally, photographic equipment was installed in high-orbiting satellites, providing better images to create improved maps.

# Combat and Reconstruction: From the Gulf War to the Global War on Terrorism

When Iraqi forces invaded Kuwait in August 1990, the United States began to assemble a military and political coalition that would ultimately drive the Iraqis out. The liberation of Kuwait was the centerpiece of Operation Desert Storm, but the coalition's accomplishments on the battlefield were predicated on a large and often overlooked logistics effort that made the offensive possible. The Corps of Engineers was a vital part of that effort, deploying 160 people to Saudi Arabia to manage the construction of nearly $300 million of base camps, sanitation facilities, roads, bridges, warehouses, and maintenance facilities. In addition, Corps real estate specialists leased hundreds of Saudi facilities, ranging from housing complexes to warehouses to maintenance facilities, to accommodate the rapidly expanding Army, Navy, and Air Force presence in the country. In addition, scientists and engineers from the Corps' research laboratories developed new technologies for analyzing terrain, detecting mines, locating water, and controlling dust that helped coalition forces operate in the harsh desert environment.

After coalition forces drove the Iraqis out of Kuwait in March 1991,

*An abandoned Iraqi tank with burning oil wells in the distance, March 1991*

Cpt. Chris Beck of the Afghan Area Office discusses the construction of the Afghan National Army facilities at Pol-e-Charki, August 2003.

the Corps of Engineers played a leading role in rebuilding the war-weary nation. Working closely with the Kuwaiti government, the Corps of Engineers established the Kuwait Emergency Recovery Office to provide project management, engineering services, and contracting support for the reconstruction effort. Over the course of the next year, the Corps helped to repair hundreds of schools and government buildings, numerous hospitals, 3,000 miles of 300-kilovolt power lines, ninety electrical substations, water and sanitation systems, the international airport and two military airfields, 150 miles of national highways, eight bridges, and two deep-water shipping ports. The Corps also supervised the construction of Camp Doha, a base for 5,000 U.S. troops that were subsequently stationed in Kuwait.

As a part of its reconstruction efforts the Corps also engaged in the largest oil-fire-fighting campaign in history. When Iraqi soldiers withdrew from Kuwait they set fire to more than 600 oil wells. The result was devastating, an environmental catastrophe that darkened the skies over Kuwait with billowing clouds of smoke, leaving huge pools of oil on the desert surface. Capping the wells and bringing the fires under control was an intensive effort, but the last of the wells was sealed off in November 1991.

For a decade after the Gulf War, the United States maintained an uneasy relationship with the nations of Southwest Asia, attempting to unsuccessfully broker some type of lasting peace in the region. The continuing unrest in the region touched the United States on September 11, 2001, when terrorists launched devastating attacks on New York City and Washington, D.C. When the Taliban regime in Afghanistan refused to expel the al Qaeda elements that planned the attacks of September 11th, the United States took military action. The United States and its Afghan allies began offensive operations in October, and by early December 2001 forced the Taliban government out of power. In the months that followed the United

States and its coalition allies helped the Afghans form a new government and a new Afghan National Army. In October 2002 the Corps of Engineers established the Afghan Area Office (AAO) in Kabul to build barracks and facilities for the fledgling Afghan army. The office also provided construction management for a variety of U.S. Agency for International Development projects in Afghanistan including the construction of roads, bridges, schools, and medical clinics. The AAO also provided engineering support for U.S and coalition forces in Afghanistan and throughout central Asia. In recognition of the office's expanded workload, in the spring of 2004 the Corps of Engineers established the Afghan Engineer District in Kabul.

When the Global War on Terrorism expanded to Iraq, the Corps of Engineers participated in pre-war planning prior to the invasion of that country in March 2003. Shortly before the war, Corps planners helped prepare a database of Iraq's transportation, oil, and electrical infrastructure and after the air war began they helped prepare target lists and advised coalition forces on targeting decisions. At the outset of the war, Corps of Engineers personnel, operating in close coordination with ground forces, helped capture and secure Iraq's southern and northern oil fields. In the southern oil fields the Corps of Engineers' Task Force Restore Iraqi Oil (TF RIO) and its contractors were instrumental in extinguishing the oil well

*Barracks take shape at the Afghan army's Central Corps headquarters at Pol-e-Charki. Building the army facilities was the first major construction program in Afghanistan in decades, August 2003.*

Protected by a heat shield and streams of water, firefighters approach a burning oil well, April 2003.

fires set by the retreating Iraqis. Combat engineers such as the 249th Engineer Battalion participated in the capture of hydroelectric facilities at the Haditha Dam and later helped the dam's Iraqi staff resume electricity production.

Sgt. First Class Paul Ray Smith

An Army engineer also became the first recipient of the Medal of Honor in Iraq. Sergeant First Class Paul Ray Smith served with the 11th Engineer Battalion, 3rd Infantry Division. On the evening of April 4, 2003, his unit was attacked by Republican Guard troops near the Baghdad airport. To hold off the company-sized enemy force, Smith climbed aboard a damaged armored personnel carrier and repulsed the enemy attack using the vehicle's .50 caliber machine gun. Sergeant First Class Smith was mortally wounded during the engagement. For single-handedly saving the lives of his men and by killing at least half of the opposing enemy force, Smith was posthumously awarded the Nation's highest award for valor.

Soon after U.S. forces toppled the regime of Saddam Hussein, the Corps

*Engineers inspect the construction at logistics support area (LSA) Anaconda, July 2005*

of Engineers began to address two vital concerns—helping the Iraqis resume the production of oil and jump starting the nation's battered electrical infrastructure. To revamp the Iraqi oil infrastructure, Task Force Restore Iraqi Oil (TF RIO) began to repair worn or damaged facilities including oil pipelines, pumping stations, gas-oil separation plants, and refineries. Immediately after the war, when Iraq was neither pumping nor refining oil for domestic consumption, TF RIO also was in charge of importing hundreds of millions of gallons of benzene and diesel fuel, and hundreds of thousands of tons of liquid petroleum gas to sustain the country.

In the fall of 2003 the Corps of Engineers established Task Force Restore Iraqi Electricity (TF RIE) to bolster electrical production and enhance the distribution of power throughout the country. Working closely with their Iraqi counterparts, RIE engineers helped refurbish Iraqi power plants, build new generating capacity, rebuild hundreds of miles of electrical transmission lines, construct new electrical substations, and install automated control systems to monitor the flow of power across the nation's electrical grid.

But the rehabilitation of the Iraqi oil and electrical infrastructure was only part of a much larger effort by the American-led coalition to help rebuild Iraq and create a safe, stable, and secure nation. Toward that end, through the Iraq Relief and Reconstruction Fund, the U.S.

*Engineers inspect the construction of a border fort near Kirkuk, May 2005.*

government allocated approximately $11 billion for 3,000 reconstruction projects that included the construction or rehabilitation of Iraq's transportation facilities, water and sewage treatment plants, hospitals and local health clinics, schools, fire and police stations, and border forts. To provide construction management for the huge undertaking, as well as provide military construction and maintenance services for the U.S. military, in January 2004 the Corps of Engineers established the Gulf Region Division (GRD). Headquartered in Baghdad, the division encompassed three engineer districts located in the southern, central, and northern parts of the country. GRD was staffed with approximately 500 civilians and 200 military personnel. All of the civilians were volunteers, and operations in Iraq marked the first time the Corps of Engineers sent such a large contingent of civilians into a combat zone.

In addition to reconstruction, the Gulf Region Division also was responsible for conducting a wide range of military construction projects in support of coalition forces operating in Iraq. Other Corps of Engineers missions in that country included collecting 600,000 tons of Iraqi ordnance from arms caches scattered around the country, destroying the unusable munitions, and storing the rest in secure depots for use by the new Iraqi army. The Corps also deployed archeologists to Iraq to help with the somber task of exhuming the bodies of thousands of Iraqis murdered by the former regime.

A key component of the Corps of Engineers' operations in Iraq was the administrative and technical support provided by Corps employees based in the United States and Europe. Another important element of GRD's success was the ever increasing role played by its Iraqi employees. The division employed several hundred Iraqis who served in a wide variety of professional and support functions.

Indeed, training the Iraqis to enhance their technical and managerial skills has been an important part of GRD's overall mission. Training host nation personnel has been an important element of the Corps' overseas programs since the Second World War.

Since 1990 the Corps of Engineers has participated in combat operations in the Gulf War and again in Iraq in 2003; in both cases those operations proved to be only a prelude to the massive reconstruction activities that followed. Through its reconstruction activities, the Corps of Engineers has played a vital role in helping Kuwait, Afghanistan, and Iraq begin the difficult and uncertain process of emerging from the turmoil of war.

# The Reconstruction of Kuwait

On February 28, 1991, a cease-fire ended military operations in the Gulf War. After a 100-hour-long ground offensive, coalition forces had achieved their objective: Iraqi forces had been forced out of Kuwait and the small Gulf nation was liberated. But the end of combat operations yielded a host of new challenges. When Iraqi forces withdrew from Kuwait they left much of the country in ruins. Consequently, at the end of the war, the U.S. Army Corps of Engineers' mission rapidly transitioned from one of supporting military operations to helping the people of Kuwait rebuild their battered country.

The Corps' role in the reconstruction of Kuwait actually began long before coalition forces took the offensive. Anticipating the destruction that could accompany the liberation of their country, in October 1990 the Kuwaiti government requested the Department of Defense's help in rebuilding their country after the cessation of hostilities. As a result of those overtures, on November 20, 1990, the Army Staff directed the Corps of Engineers to serve as the lead agent in assisting the Kuwaiti government to rebuild its public works and municipal utilities.

In January 1991, Chief of Engineers Lieutenant General Henry Hatch directed Colonel Ralph Locurcio, the commander of the Corps' Savannah District, to establish an area office in Kuwait to oversee the reconstruction effort. That organization, which later became the Kuwait Emergency Recovery Office (KERO), was organized much like a Corps district, with separate offices for project management, emergency operations, engineering services, and contracting and support. In planning KERO operations Colonel Locurcio drew heavily on the Corps' long experience in restoring power and water supplies after natural disasters. The recovery office was staffed largely with civilian volunteers from the Corps of Engineers, many of whom had previous emergency operations experience.

The KERO advance team traveled to Saudi Arabia at the end of January and quickly procured sufficient food, water, equipment, and vehicles to sustain the office for thirty days. On March 4th, just days after the ceasefire took effect, the first KERO personnel arrived in Kuwait City. They found the city in shambles. There was no electricity, the municipal water and sanitation systems had been destroyed by the retreating Iraqis, and thousands of

*Abandoned Iraqi vehicles litter the highway heading north out of Kuwait City*

burned out Iraqi tanks and abandoned vehicles littered the streets.

KERO was initially attached to Task Force Freedom, the Army's coordinating activity for the reconstruction of Kuwait. Within hours of arriving in Kuwait City, KERO engineers, assisted by Kuwaiti volunteers, began fanning across the city to conduct damage assessments. The KERO damage assessment groups inspected ports, the Kuwait airport, the wastewater treatment system, power production and distribution facilities, public buildings, and defense installations. During its first forty-five days of the operation, KERO teams conducted more than 1,000 assessments that served as the foundation for later reconstruction efforts, many of which were managed by the Corps of Engineers. KERO expanded along with its workload, and by the end of March had a staff of 14 military officers, 112 Corps civilians, more than sixty Kuwaiti volunteers, and nearly 1,000 contractors.

KERO was a key member of a U.S. Army effort that quickly restored Kuwait's primary power systems within thirty days, replenished the nation's water supplies, and reopened the badly damaged airport within forty-five days. KERO's largest single mission was the restoration of Kuwait's public buildings. Working together, KERO and its contractors restored more than 1,000 public buildings including 145 schools, the Kuwait Airport, and the National Assembly building. By December 1991, a scant nine months after the end of the war, KERO had restored power to 99 percent of the country, returned three desalinization plants to operation, reconstructed two sewage treatment facilities, and completed an assessment of the entire sanitary system. The rehabilitation of the Kuwait transportation system also included repairs to more than 150 miles of road, and the removal of 3,700 bunkers, barriers, and abandoned or destroyed vehicles.

"The contribution of the U.S. Army Corps of Engineers to the reconstruction of Kuwait is a source of pride to the entire U.S. mission," wrote Ambassador Edward Gnehm in a letter to Colonel Charles Cox. "The achievements of your engineers have won high praise from both the government of Kuwait and its people." On another level, the working relationships forged between the Kuwaiti government and the Corps of Engineers during the reconstruction served both countries well when the United States traveled back to Southwest Asia in early 2003 to begin combat operations against Iraq.

# 4 A Vital Part of the Army

The Pentagon under construction, 1942

# Military Construction

The military construction mission of the U.S. Army Corps of Engineers dates from just prior to World War II. Until that time, the Quartermaster Department built almost all facilities for the U.S. Army. By 1940, it was clear that this arrangement could not continue. Quartermaster resources were inadequate for the large mobilization job ahead. Furthermore, the engineers' civil works organization and experience provided the basis for absorption of the new assignment. So in November 1940, the War Department chose the Corps to build facilities for the Army Air Corps. Thirteen months later, the Corps of Engineers undertook all construction for the U.S. Army's war effort.

This massive enterprise involved military and industrial projects. The Corps managed construction of a wide range of factories, most notably for the assembly of aircraft and tanks and the production of ammunition. Corps-built military installations included camps for 5.3 million Soldiers, depots, ports, and the Pentagon. Each of these tasks included planning, site selection, land acquisition, design, contract negotiations, procurement, labor relations, and the construction itself. All told, the wartime mobilization program involved more than 27,000 projects and cost $15.3 billion. Major General Leslie R. Groves, head of the Manhattan Project, summed up the significance of this work for the successful conduct of the war: "Mobilization was decisive and construction generally controlled mobilization."

Yet there was more to U.S. Army engineer construction during the war than the stateside program. Work in

*Hangar under construction by the Jacksonville District at MacDill Field, Tampa, Fla., for Army Air Corps, January 1942*

*Soldiers of the 95th Engineer General Service Regiment, an African-American unit, building a bridge on the Alaska-Canada (ALCAN) Highway.*

*Steel barge bridge along the Ledo Road*

support of the war against Japan ranged over a vast portion of the world, from Panama to India and from Alaska to Australia. A huge organization—which grew to include 236,000 engineer troops in an Army of 1,455,000—built pipelines, dredged harbors, and built and repaired ports throughout the Pacific Theater.

The accomplishments in the Pacific rivaled those of the Corps on the home front. Among the major projects in the Pacific area was the air ferry route to the Philippines. To move heavy bombers west across the ocean, the Corps built airfields on a host of Pacific islands. U.S. Army engineers developed these bases in a matter of a few months.

Two land routes also merit special notice. The ALCAN Highway, from Dawson Creek, British Columbia, to Fairbanks, Alaska, prompted by the threat of a Japanese invasion and the closure of Alaskan sea routes, ran through nearly 1,600 miles of muskeg and mountains. The project, begun in 1942, involved 133 major bridges and, at the peak of construction, employed eighty-one contractors and 14,000 men. Closer to hostilities, the Ledo Road from northeastern India to Burma crossed 430 miles of jungle, mountains, and rivers. Paralleling the road was the longest invasion pipeline ever built. Construction began under difficult conditions in

Fitzsimmons Army Hospital, Denver, Colo., 1952

late 1942 and was completed when a convoy from India reached China in early 1945.

The war against Germany also demanded massive construction support. After building bases in Greenland and Iceland to protect Atlantic shipping, the Corps moved to England, where as many as 61,000 U.S. Army engineers created the ground and air facilities required to support the strategic bombing of Germany and the invasion of France. During the same period, in North Africa the Corps built many airfields for British and American air forces and provided ports and depots to support the invasion of Italy.

In June 1944, engineers moved into Europe with the Allied invasion. Operations included the rehabilitation of ports and railroads as well as airfield and depot construction. For example, engineers cleared and reconstructed the port of Le Havre using plans developed well before the advance into France. Large construction projects also included a camp and depot at Valognes, France, that served as headquarters for logistical forces of the Communications Zone. The post included tents for 11,000 Soldiers and provided 560,000 square feet of hutted office space.

After the war, the Corps maintained a large presence in Europe. Engineers restored transportation networks and other public services in Germany and Austria. In France during the early 1950s, the Corps performed a wide array of line-of-communications construction, from pipelines to supply depots, in anticipation of the need to reinforce units in Germany. Additionally, U.S. Army engineer construction fulfilled the needs of the large numbers of

American troops stationed in Germany through the end of the Cold War by building housing, hospitals, depots, and offices.

The U.S. Army Corps of Engineers also remained with the occupation forces in Japan and met all of their building requirements. When war broke out in Korea in 1950, bases in Japan provided the springboard for the movement and supply of forces deployed against the North Koreans and Chinese. In Korea itself, engineers performed remarkable feats of road and bridge construction over extremely difficult terrain and provided ports and airfields for friendly forces. They rehabilitated water supply and sanitation systems that remained in use by the Republic of Korea for many years, and they still provide construction support for American units stationed there.

Military construction after the Korean War expanded into numerous countries. Work continued in Europe and the Far East, but increasing Cold War tensions led to the establishment of bases elsewhere. Through the 1950s and into the 1960s, the Corps built early warning facilities and airbases in diverse locales, including Greenland, Morocco, and

*Buildings constructed by the Corps on Greenland to support Project Blue-Jay, October 1951. Project Blue-Jay involved the construction of Thule Air Force Base.*

*Air Force Dormitory, Brindisi, Italy, 1965*

Libya. These forward bases brought Strategic Air Command bombers within striking range of the Soviet Union.

After the Soviet Union tested its first atomic bomb in August 1949, the United States began looking for ways to protect its vital military installations and major urban areas from Soviet air attack. The answer was the U.S. Army's Nike antiaircraft missile system, and in 1952 the U.S. Army Corps of Engineers began purchasing land and building Nike missile batteries at sites around the country. Each site encompassed approximately forty acres, and between 1954 and 1958 the Corps built nearly two hundred Nike Ajax missile batteries. In 1958 the Army began replacing the liquid-fuel Ajax missiles with the longer-range, solid-fuel Nike Hercules equipped with nuclear warheads. To house the new missiles the Corps of Engineers either modified the existing Ajax facilities or built new Nike Hercules missile batteries. Ultimately the Corps of Engineers constructed a total of 265 Nike Ajax and Hercules launch facilities. The last Ajax battery was decommissioned in 1963 and the final Hercules missile site was closed in 1979.

Even as the United States was building an air defense network, the evolution of a new technology—long-range intercontinental ballistic missiles (ICBM) armed with nuclear weapons—opened a new chapter in the arms race with the Soviet Union. While the United States Air Force raced to develop an operational ICBM, in 1957 it turned to the U.S. Army Corps of Engineers to begin building the research, test, and training facilities to support the development effort, as well as the operational launch sites to deploy the ICBMs. In 1960 the Corps established the Corps of

*The ICBM site construction program spanned the country and encompassed facilities from New York to California.*

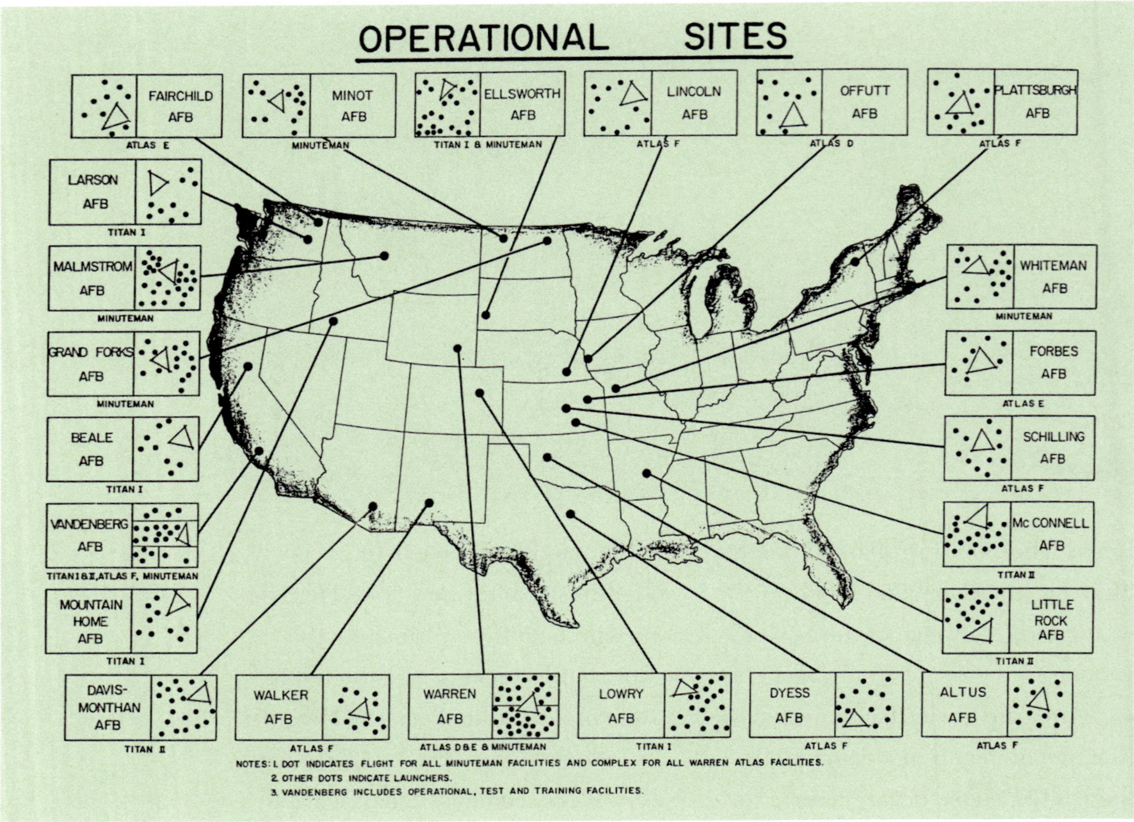

*A Nike Ajax missile battery in 1959. The heavy earthen berm on the right surrounded the refueling area.*

*National Air and Space Museum*

Engineers Ballistic Missile Construction Office (CEBMCO) to manage the project. By 1966 CEBMCO had a staff of three thousand people managing twenty-two construction projects spread over seventeen states. Construction of the missile facilities went on around the clock, and by 1961 more than twenty-one thousand construction workers were building missile facilities. Construction of the Atlas, Titan, and Minuteman missile silos, most of which were built deep underground and hardened to survive a preemptive first strike, was particularly challenging and required the Corps to develop new construction techniques and management procedures to support the effort. By the late 1960s, the Corps had completed 1,200 ICBM launch sites.

In the 1970s the Corps provided construction support for the Sentinel and Safeguard antiballistic missile (ABM) programs. The ABM construction program culminated in the completion of the Stanley R. Mickelsen Safeguard Complex in North Dakota in 1972.

During the military buildup of the 1980s, the U.S. Army Corps of Engineers conducted large construction programs for the U.S. Army and the U.S. Air Force. During the first half of the decade, the construction effort reached approximately $1 billion a year for each service. In the largest U.S. Army installation construction

*The main tunnel of a Titan I launch complex nearing completion, 1960*

*Activation of the first Titan I squadron at Lowry AFB, Colo., April 1962*

*U.S. Strategic Command History Office*

*Shopping center under construction at Fort Drum, N.Y.*

program since World War II, the Corps built an almost completely new base at Fort Drum, New York, for a newly organized light infantry division, the 10th Mountain. Although the division used some of the existing buildings, the Corps constructed almost an entirely new post, including infrastructure, barracks, family housing, dining facilities, headquarters buildings, a large physical fitness complex, medical clinics, and an airfield. Built on a tight schedule, the almost $1 billion construction program produced a modern, well-planned installation adapted to its environment and incorporating lessons learned at other U.S. Army installations. With its enclosed shopping mall, child care center,

*Barracks under construction, Vilseck, West Germany, 1983*

and recreational and entertainment facilities, the installation reflected the U.S. Army's growing concern about the quality of life of its Soldiers and their families. Although unique in its scope and complexity, the Fort Drum program was only one portion of the busy Army and Air Force construction programs of the Reagan administration.

With the collapse of the Soviet Union and the end of the Cold War, the future of military construction was uncertain. Many military construction projects were temporarily frozen as the Nation's leaders discussed the possibility of a "peace dividend." As the military services struggled to redefine themselves in the post-Cold War world, the Army began to consolidate installations and dispose of unneeded property. The Base Realignment and Closure (BRAC) program was an attempt to save money and adapt the installation structure to the expected decline in the services' size. BRAC, however, generated its own demand for construction, as units moved to new installations that required new facilities.

The U.S. Army Corps of Engineers was also active in the effort mandated by international convention to dispose of chemical weapons that were outdated or no longer needed in the Nation's arsenal of weapons. The Chemical Demilitarization Program involved the construction of complex and expensive facilities that, although at times controversial, were designed to dispose of the chemical weapons located at eight sites within the Continental United States and one on Johnston Atoll in the Pacific Ocean.

The Department of Defense began an ambitious environmental cleanup program in 1984. At former and current sites, the services worked to locate and remove old contaminants and operate active installations in an environmentally responsible manner. Much of the work associated with these programs fell to the U.S. Army Corps of Engineers. In 1997, the Corps' environmental cleanup duties expanded when the Formerly Utilized Sites Remedial Action Program (FUSRAP) was transferred from the Department of Energy to

*Preparing to remove underground storage tank at the former Kincheloe Air Force Base, Kinross, Mich., 1994*

*Barracks at Fort Bragg, N.C., 2003*

the Corps. FUSRAP removed radioactive materials from sites formerly used by the Manhattan Engineer District, which built the Nation's first nuclear weapons during World War II, and its successor, the Atomic Energy Commission.

As part of its military construction mission, the Corps continued to have responsibility for the renovation of the Pentagon, a structure that it had built during World War II. Nearly six decades later, the Pentagon badly needed repair and updating. The Corps completed the first segment of the renovation before responsibility for the massive renovation project was transferred to another agency in 2000. The Corps' work proved its durability when it resisted the impact of the September 11, 2001, terrorist airliner attack much better than the adjacent, unrenovated segment of the building.

Other military construction programs aimed to improve the quality of life for Soldiers. A major barracks renovation program provided better facilities with more amenities and privacy to enlisted Soldiers, and a massive new housing privatization program began placing large proportions of U.S. Army family housing in the hands of private companies. Under the Residential Communities Initiative, contractors began renovating and improving existing family housing and building large tracts of new housing. The Nation's reliance on an all-volunteer Army meant that the quality of life for Soldiers—who were increasingly deployed in combat abroad—and their families at home was an important priority.

Even before the terrorist attacks of 2001, it had become apparent that the post-Cold War world would not be a peaceful one. After years of research and development, the United States began acquiring weapons and building facilities to provide a defense against a limited ballistic missile attack, and the U.S. Army Corps of Engineers played an important role in providing the ground-based facilities in Alaska. But increasingly, the country found itself drawn into smaller conflicts like the civil strife that plagued Somalia, Rwanda, and the collapsing Yugoslavia. Large and rigid Cold War-era U.S. Army units were difficult to use in this new combat environment, and in 1999 Chief of Staff of the Army General Eric Shinseki began a massive reorganization of combat units to make them smaller, lighter, and more flexible. The Corps helped to design and build the new bases that would train and support these new units.

U.S. Army transformation led to "Milcon Transformation" with the objective of providing these new facilities faster, better, and cheaper in close cooperation with private industry. One of the early challenges was to provide modular facilities quickly for troops who were moving back to the United States from Iraq and other parts of the world and preparing for transformation.

In the early years of the twenty-first century, the Corps confronted challenges inherent in executing its normal military construction mission for the Army, the Air Force, and other Department of Defense agencies;

*Exterior view of Camp Zama, Japan, high rise family housing, 1999*

supporting the massive spending on the Global War on Terrorism in Iraq and Afghanistan; supporting Army Transformation; and preparing for an additional round of BRAC requirements. Although the Cold War with its large demands on the Corps had ended, the post-Cold War world offered a new and daunting set of challenges that were scarcely imagined just a decade earlier.

# The U.S. Army Corps of Engineers Responded to President Kennedy's Call for National Preparedness

"We are not against any man—or any nation—or any system—except as it is hostile to freedom." So stated President John F. Kennedy in a May 25, 1961, special address to Congress on urgent national needs in response to crises in Berlin, Germany, and Cuba. In the address, President Kennedy spoke at length on civil defense, which he characterized as "insurance for the civilian population in case of an enemy miscalculation." To overcome years of neglect, he assigned responsibility for civil defense to Secretary of Defense Robert McNamara and established a National Fallout Shelter Program.

Secretary McNamara proceeded to create an Office of Civil Defense within the Department of Defense and tapped the U.S. Army Corps of Engineers and the Navy's Bureau of Yards and Docks to conduct a fallout shelter survey and other civil defense tasks. The initial mission was to identify structures, determine their ability to block a massive dosage of radiation resulting from a nuclear attack by a factor of twenty, and mark them as public shelters. The goal was to find shelter for up to 50 million Americans.

The Corps responded by creating a Joint Civil Defense Support Group in the Chief's office with a colonel in charge. The Corps staffed the new headquarters organization and similar offices with division and district personnel. Most of these personnel were diverted from civil works assignments.

Within a short time, the National Fallout Shelter Survey achieved impressive results. The Corps developed specialized techniques for computer processing of survey data, developed scientific methods to evaluate potential shelters, trained nearly 1,500 architect-engineers and Corps employees, and negotiated and supervised more than 500 architectural and engineering contracts to conduct the nationwide survey. The fallout shelters thus established were stocked with federally procured water, food, medical, and sanitation supplies, as well as radiation monitoring kits.

Additional civil defense tasks included preparing the following: engineering and cost studies of standard structures for emergency operating headquarters, pilot feasibility studies to determine local capabilities to quickly increase the number of public shelters, technical civil defense publications, a nationwide survey of construction and engineering equipment and inventory of potential contractors, and a survey of fallout shelters for selected radio and television stations in the National Emergency Broadcast Network.

The program continued throughout the 1960s, and by 1970 it was consolidated at the Corps' division level. Overall management passed to the Defense Civil Preparedness Agency in 1972. This organization was subsumed into the Federal Emergency Management Agency in 1979.

The Corps' response to President Kennedy's call for national preparedness was another example of the agency's ability to quickly and efficiently respond to new missions using its decentralized organization and established contracting expertise.

*Precast concrete plant,
King Khalid Military City,
Saudi Arabia*

# Work for Other Nations

Shortly after World War II, the U.S. Army Corps of Engineers became involved in massive foreign assistance programs sponsored by the United States government in response to the devastating impacts of that global conflict. Much of Europe was a shambles, suffering in many instances from physical devastation and political instability. These conditions made the continent vulnerable to the expansion goals of the Soviet Union. As a result, in 1948 the U.S. Congress approved Secretary of State George C. Marshall's plan to provide financial support for reconstruction programs developed by participating European nations. This ambitious plan followed separate congressional aid packages to Greece and Turkey, nations that were particularly vulnerable to subversion or aggression.

The 1951 Mutual Security Act extended the U.S. foreign assistance program to other portions of the globe. This law was passed in a period of growing international tensions marked by the advent of the Iron Curtain, the Berlin Blockade, the communist success in China, and the outbreak of the Korean War. The purpose of the legislation was maintenance of national security and promotion of U.S. foreign policy through military, economic, and technical assistance to strengthen friendly nations. The act consolidated or built upon a variety of efforts, including the Military Assistance Program authorized in 1949 by the Mutual Defense Assistance Act, through which the United States offered help to allies in establishing defenses against external aggression and internal violence. The Mutual Security Act also included the program of technical assistance first articulated in President Harry S. Truman's 1949 inaugural address. Finally, the new law replaced the various economic aid programs with comprehensive loan and grant provisions.

Foreign assistance programs continued to evolve in response to changing perceptions of the world situation and American interests. In the first years of the Cold War, economic aid predominated. During the Eisenhower years, from 1953 through 1961, most of the assistance

*Gen. George C. Marshall*

*Interior of Dhahran Airport, Saudi Arabia, completed by the Corps in 1961*

from the United States was military. Then, in the decade that followed, an equilibrium was reached between economic assistance and military programs, including sales. The Foreign Assistance Act of 1961 established the U.S. Agency for International Development (AID) to administer the major economic aid programs. More significantly for later U.S. Army Corps of Engineers activities, Section 607 of this act provided for furnishing services and commodities to foreign countries on a reimbursable basis. Starting in the mid-1960s, this became the basis for a number of major engineering programs.

Other important trends shaped the role of the U.S. Army Corps of Engineers abroad. As bipolar hostilities appeared outside of Europe, base construction spread from Middle Eastern and North African countries to the Far East and South Asia. This trend coincided with the advent of a different form for transferring aid to recipient nations. During the early years of the Cold War, most aid was in the form of grants—90 percent of American help was outright gifts. By the mid-1960s, 60 percent of economic aid was loans.

The U.S. Army Corps of Engineers' contributions to these foreign programs took place in this context of evolving emphasis. Thus, during the immediate postwar years, when American foreign policy and assistance programs emphasized Europe and particularly Greece and Turkey, the Corps was extremely active in these two nations. In Turkey, the Corps concentrated on construction of military facilities for Turkish and American armed forces. In Greece, after the State Department came to the Corps for technical expertise, the Corps restored a badly mauled transportation and communication network. The Grecian District, which was established in Athens in July 1947, cleared the Corinth Canal, restored the Port of Piraeus, and built or repaired more than 3,000 kilometers of roads.

Corps operations in Greece established several major precedents. First was the organization of an engineer district to administer and supervise large-scale infrastructure programs in a foreign country. Second was the provision of technical assistance in conjunction with economic aid. Third, the practice of training local contractors and artisans to perform as much of the actual work as possible became an integral part of reconstruction and economic development. Fourth, the commitment to helping a friendly nation to help itself, which was manifested in projects aimed at restoring the Greek economy, became a standard feature of Corps projects.

*Reconstruction of the port of Piraeus, Greece, February 1948*

During the 1950s, the Military Assistance Program dominated American overseas efforts. This program was one of two major Department of Defense foreign activities in which the Corps participated. First and most important was the maintenance and support of American forces in other lands. The other, the Military Assistance Program through which the United States aided the military forces of other nations, was directed largely toward supporting allies on the periphery of the Soviet Union and near the People's Republic of China.

In the period 1950–1964, this program dispensed assistance valued at more than $350 million. Iran, which was the largest single recipient,

The Corps built a cantonment for the Pakistani army at Multan to accommodate 8,500 troops. The barracks above were two structures of the 310 built on the post. Construction at Multan began in 1959 and was complete two years later.

and four other nations—Pakistan, Turkey, Taiwan, and South Korea—received nearly all of the military assistance money. The projects carried out in Pakistan by the Trans-East District of the Mediterranean Division illustrate the nature of the work performed. In a massive modernization program for the Pakistani armed forces, the Corps built cantonments, airfields, wharves, and marine railways.

While heavily involved in these efforts, the U.S. Army Corps of Engineers also worked in programs of economic assistance. Projects intended to buttress a recipient nation's economy were administered by AID and its predecessor agencies. Corps participation in economic development programs actually predated the establishment of any of these agencies. As early as 1946, the Corps of Engineers worked with numerous Latin American governments to establish national cartographic programs. These efforts were ultimately intended to provide the basis for resource inventories of participating nations. After 1953, when the Department of State took over this program, the Corps continued to contribute to its success. Engineer personnel worked in twenty-two countries developing programs, rendering procurement assistance, and administering contracts.

The Corps of Engineers' Peshawar, Pakistan, resident office opened in 1956 in this remote city near the Khyber Pass. The resident office first supervised construction for the Pakistani air force, and then in 1958 worked on highly classified projects for the U.S. Air Force. Shortly after war broke out between India and Pakistan in 1965, Air Force work ended and the office closed.

In the late 1950s, the Corps began undertaking large projects within the economic assistance program. Between 1950 and 1964, the

Corps produced major engineering studies for seventeen different countries. These surveys dealt with beach erosion problems, river hydraulics, transportation networks, and entire public works programs. Corps personnel examined the feasibility of various port and highway projects. Engineers also became involved in actual construction in eight countries. The major construction projects included airports, highway systems, and ports, and the Corps spent $109.5 million on them between 1959 and 1964.

The Corps' work on these studies and construction projects reflected new directions in the overall program administered by AID. In the years just prior to 1965, the focus was on long-term projects that supported broad economic development. In this framework, engineering and construction loomed large, and the Corps, with its unique capability to plan, organize, and execute major building programs, made major contributions.

During the mid-1960s, several developments led to changes in the Corps' role in foreign programs. AID changed its emphasis from major construction efforts aimed at improving economic infrastructures to more immediate needs for the improvement of food supplies, public health, and education. Moreover, AID turned more to private engineering and architectural firms for support in this area. In so doing, the agency cited the provisions of Section 601 of the Foreign Assistance Act of 1961, which encouraged maximum utilization of private resources instead of other government agencies.

The buildup of American armed forces in Vietnam also redirected the foreign operations of the U.S. Army Corps of Engineers. The maintenance and support of American forces in Southeast Asia took an ever-increasing portion of the Corps' resources. Moreover, Vietnam absorbed a growing percentage of the foreign aid budget, leaving less money for major projects in other parts of the world. As AID turned its attention to Vietnam and Southeast Asia, the agency became involved in major geodetic and cartographic enterprises. The U.S. Army Corps of

*Housing courtyard, King Abdul Aziz Military Academy, Saudi Arabia*

*Saudi Arabian National Guard headquarters building*

Engineers, with expertise already employed in a number of other nations, contributed again to resource inventory projects and the production of maps required for the land reform program of the government of South Vietnam. Thus, while the Corps' involvement in major construction projects dropped off, it still participated in other aspects of AID's work.

Even before international developments had changed the character of U.S. Army Corps of Engineers' overseas projects, another major factor had entered the picture. This was the beginning of Corps involvement in reimbursable programs funded by recipient nations instead of by U.S. loans and grants. Authorized by Section 607 of the Foreign Assistance Act, these projects were based on bilateral agreements between the United States and nations that sought the Corps' technical expertise in development programs. The first of these was funded by the government of Saudi Arabia in 1963. There the Corps engaged in a large number of construction projects—including a variety of facilities for the Saudi Arabian armed forces and civil projects such as construction of radio and television communications installations—that eventually totaled $5 billion when it ended in the late 1980s.

By the late 1960s and early 1970s, the number of reimbursable programs had grown. In addition to the work in Saudi Arabia, projects started in Iran, Jordan, Kuwait, and Libya. The Corps' effort in these nations improved the American balance of payments and provided valuable experience for U.S. Army engineering personnel while sharing the Corps' technical and professional expertise.

The U.S. Army Corps of Engineers met more pressing requirements in the Middle East while managing its long-term reimbursable projects. In accordance with the 1978 Camp David Agreements, the Corps built two airbases for Israel as replacements for those evacuated during the withdrawal from the Sinai. Completed in 1982, only three years after the start of construction, the bases cost about $1 billion, more than three-fourths of which was an American grant. Meanwhile, the Corps also constructed Sinai base camps for the multinational force and observers who patrolled the demilitarized zone between Egypt and Israel.

Egypt also received considerations as a result of the Camp David Agreement. In addition to the opportunity to obtain F-16 jets through the Peace Vector program, the Egyptian air force received improvements to airbases to accommodate these new aircraft. An example of the base improvement effort was the large Gianaklis airbase in the Nile delta, a $250 million project awarded in 1992 and substantially completed by 1996.

After the Wye River memorandum of 1998, the Corps again participated in attempts to maintain peace in the Middle East. In exchange for moving bases from the West Bank and thereby freeing land for possible transfer to the Palestinians in accordance with the Wye River memorandum, the Israelis received two infantry training bases and other facilities paid for by the United States and constructed by the Corps. Although the reimbursable programs of recent years have been less extensive than the massive Saudi Arabian and Israeli airbase projects, reimbursable work continued to be an important Corps mission.

The U.S. Army Corps of Engineers has consistently played a major supporting role in "nation building" around the world. The

*Construction of enlisted quarters, King Khalid Military City, Saudi Arabia*

*Hardened aircraft shelter with an F-16D completed by the Transatlantic Programs Center at Gianaklis Airbase, Egypt, 1996.*

Port of Owendo, Gabon, site of Corps studies undertaken for the Agency for International Development

wide variety of projects to help other nations has included technical assistance to the African nation of Gabon to improve its ports, geological and hydrological studies of the Niger River Basin in Africa, technical advice on water resources development to the People's Republic of China, disaster relief in Bangladesh after devastating floods in 1991, and construction of hydropower facilities in the Federated States of Micronesia.

The collapse of the Soviet Union and the end of the Cold War in the 1990s produced large construction programs in the former Soviet Union. Although financed by the United States, these programs responded to and reflected the new geopolitical realities in the world. The breakup of the Union of Soviet Socialist Republics led to the creation of a number of new nations that needed U.S. embassies, which the Corps helped construct or renovate. A large program began in 1997 as a result of concern about the handling of nuclear weapons in the former Soviet republics. The Cooperative Threat Reduction Program funded a variety of cooperative construction projects, ranging from the building of a Russian facility to store fissile materials from dismantled nuclear

Holding area for fissile material in Russia

weapons to the construction of apartment buildings in the Ukraine for former soldiers of the Soviet Strategic Rocket Forces who required housing. In another program in the former Soviet Union, the Corps, in cooperation with the U.S. Customs Service and the Republic of Georgia, built facilities to help the Georgian government secure its borders to inhibit the movement of dangerous cargo such as drugs or nuclear weapons and increase its customs revenues. All of these programs sought to bring some stability to a vast area undergoing the difficult transition to new political and economic systems.

Often overshadowed by such large programs are a variety of small projects that affect the lives of perhaps only a few, but with possible implications for many. The Corps has worked in more than 30 African nations on numerous small infrastructure projects like roads, bridges, schools, water wells, low-cost housing, health clinics, sanitation facilities, and biodiversity promotion. Working with U.S. embassies and local military forces, the Corps has built facilities such as a community training and counseling center for the Kenyan Red Cross to assist in its struggle with the devastating effect of HIV/AIDS and drug abuse. In addition, the Corps provided assistance to AID in the wake of the 1998 embassy bombings in Kenya and Tanzania to help mitigate damage to surrounding buildings, and a myriad of reconstruction projects following the wars in Afghanistan and Iraq.

*A Corps-built clinic, Rwanda*

Whatever the scope of the project, the U.S. Army Corps of Engineers has sought since the end of World War II to assist other nations in improving their infrastructures, to share American technical know-how, and to help other countries cultivate their own capabilities for self-development. From large-scale construction programs like the massive Saudi Arabian effort to smaller feasibility studies in the 1980s such as the harbor improvements at the Port of Asau in Western Samoa, the U.S. Army Corps of Engineers has developed the ability to assist other nations in vital engineering and construction management activities, both large and small.

# Strengthening the Free World: Rehabilitating Postwar Greece

The advantages of having a military-civilian engineer organization were demonstrated when the United States decided to help Greece recover from the devastation of war. Soon after the end of World War II, Greece was torn by a civil war. President Truman and congressional leaders believed it was in America's interest to prevent the sitting Greek government's collapse by assisting the nation to get on a path toward economic recovery. To strengthen the anticommunist monarchy, a program of economic aid to Greece was developed under the auspices of the U.S. Department of State.

President Truman appointed Dwight P. Griswold, a former governor of Nebraska, as the administrator of the recovery program. Soon after his arrival in Greece in July 1947, Griswold reported on the extensive devastation

*The dredge* Poseidon *clearing the Corinth Canal, 1947*

he found. The State Department decided that the reconstruction and rehabilitation of roads, railroads, bridges, ports, and the Corinth Canal, one of the main Greek waterways, were of primary importance. Once the country's transportation system was restored and the ports were in operable condition, economic recovery would be more rapid.

Although it received some 100 letters from construction firms interested in doing the work, the State Department was unfamiliar with doing construction and letting contracts; it had no organization to do the job. It repeatedly sent representatives to the Office of the Chief of Engineers to get information regarding such matters as the selection of contractors, the types of contracts that could be used, and the amount of the fee to be paid. The State Department concluded it would be unable to do the work itself and asked the U.S. Army Corps of Engineers, which had a capable civil works construction organization, to undertake the work on its behalf. Assigned to the Corps in late July 1947, the program was scheduled to be completed within a year.

The Corps of Engineers subsequently set up the Grecian District, headquartered in Athens, to manage the program. Its personnel were largely drawn from divisions and districts throughout the Corps. The new district entered into agreements with a number of contractors that formed joint ventures. By mid-August 1947, Colonel David W. Griffiths, the new District Engineer, some of his civilian employees, and some of the contractors' employees arrived in Athens.

Actual reconstruction began in mid-September with the clearing of debris from the port of Piraeus. Soon work was under way on the reconstruction of other ports, the repair of wrecked railroad bridges and tunnels, and the upgrading of highways, all of which had deteriorated badly. Debris-clearing operations began on the Corinth Canal. Soon after arriving in Greece, Colonel Griffiths was given the additional duty of upgrading a number of airfields.

All of this work had to be done rapidly and efficiently. Secretary of War Kenneth Royall had admonished that "the War Department is on continual exhibition to the President, the Congress, the State Department, and to Greece ... and other interested nations." Colonel George W. Marvin, the chief engineer of the American military assistance group advising the Greek Army in its fight against the guerrillas, helped Colonel Griffiths by obtaining Greek Army units to provide security for men working on District projects.

The U.S. Army Corps of Engineers reconstructed about 900 miles of highway, rebuilt three major ports, restored railroad bridges and tunnels totaling some two miles, and upgraded ten airfields. The Corinth Canal was reopened after about one million cubic yards of earth and debris had been removed. Actual construction time was about a year and a half. The schedule overrun resulted mainly from guerrilla attacks, unusually severe winter weather, and unexpected delays in getting supplies. Once again, the dual military and civilian organization of the U.S. Army Corps of Engineers made possible the efficient accomplishment of an important strategic mission.

*Greenland Ice Sheet Project (GISP), Deep Drilling Operation by the Corps' Cold Regions Research and Engineering Laboratory, 1975. Cold War strategic interest in cold regions prompted extensive research.*

# Changing Military Responsibilities and Relationships

During World War II, the Office of the Chief of Engineers and its subordinate activities exercised a broad range of military responsibilities. The Corps trained engineer officers and enlisted men, primarily at Fort Belvoir, Virginia, home of the U.S. Army's Engineer School since 1919, and at Fort Leonard Wood, Missouri, where an Engineer Replacement Training Center opened in 1941. The Corps developed the Tables of Organization and Equipment that structured U.S. Army engineer units, wrote the technical manuals that explained the use of engineer equipment, and prepared the field manuals that detailed military engineering tactics and doctrine. The Corps determined the U.S. Army's engineer equipment requirements, purchased the items needed and distributed them, while supervising the efforts of the Engineer Board to develop new and improved equipment. It

*Company barracks, Fort Belvoir, Va., home of the U.S. Army Engineer School until 1988.*

*Engineer Replacement Training Center, Fort Belvoir, Va., which also trained enlisted personnel during World War II.*

*Nuclear reactor built at Fort Belvoir, Va., in 1955–57 by the Army Engineers Reactors Group and the Atomic Energy Commission. The Army's first nuclear reactor, this facility was decommissioned in 1973.*

selected engineer officers for assignment to troop units, schools, and civil works. The Corps supervised all U.S. Army mapmaking. Finally, the engineers met the huge military construction and real estate needs of a rapidly expanding U.S. Army.

These functions, with the exception of general military construction and Army real estate, transferred to the Corps in December 1941, were traditional Corps missions that the engineers pursued during the war on a vastly expanded scale. Three months after the attack on Pearl Harbor, however, its position within the War Department changed, as the Corps of Engineers and other technical and administrative services of the U.S. Army were placed under the Services of Supply, one of three major components into which the War Department was then divided. General Brehon Somervell, himself an engineer officer, commanded this organization throughout the war, although its title changed in 1943 to Army Service Forces.

When the Army Service Forces headquarters was dissolved in 1946, the Chief of Engineers and the chiefs of the U.S. Army's other technical services returned briefly to the direct supervision of the Army chief of staff. The director of Logistics, however, inherited the general supervision of the technical services in 1948, and the deputy chief of staff for Logistics obtained more effective oversight of their work in 1954. The Under Secretary of the Army (during 1950–1953) and Assistant Secretaries of the Army for Materiel; Financial Management; Civil-Military Affairs; and Manpower, Personnel, and Reserve

Forces (during the Eisenhower administration) successively provided civilian direction for the Corps' military construction, housing, and real property functions.

For a decade and a half after World War II, the U.S. Army Corps of Engineers undertook the same broad range of functions it had exercised during the war. It even retained its role as engineering and construction agent for the U.S. Air Force after that service became independent of the U.S. Army in 1947. In 1954, the Corps became responsible for the Army's nuclear reactor program. It created the Army Engineer Reactors Group, which, in conjunction with the Atomic Energy Commission, completed in 1957 the Nation's first military nuclear power plant built primarily to generate electricity. Other nuclear plants followed, including a floating power plant and field reactors producing both steam heat and electricity.

## Research Laboratories

The Corps' laboratories prospered in the postwar years. The Engineer Research and Development Laboratories at Fort Belvoir, successor to the Engineer Board, continued its work in developing new and improved bridging, road construction, camouflage, demolition, mapping, and mechanical equipment. A Nuclear Power Branch was added to the laboratory to engage in research and development in the nuclear power field.

*Before computer modeling, the Waterways Experiment Station in Vicksburg, Miss., used physical models to study waterways problems. This model of the Mississippi River at St. Louis, Mo., tested various measures to reduce siltation in the harbor.*

A joint logistics over-the-shore sand grid demonstration test, Fort Story, Va.

The Waterways Experiment Station, established by the Corps and its Mississippi River Commission in 1929 at Vicksburg, Mississippi, as a hydraulics laboratory, had entered the field of military research and development during World War II. Soon after it developed the pierced-steel plank and prefabricated bituminous surface used in U.S. Army airfield construction. Placed under the direct supervision of the Chief of Engineers in 1949, during the Cold War the Waterways Experiment Station developed flexible pavements for runways designed for heavy B-52 bombers, and it examined, through chemical simulation, the blast effects of nuclear detonations in an effort to produce hardened

The Cold Regions Research and Engineering Laboratory, Hanover, N.H.

*A Distant Early Warning (DEW) Line station on the Greenland icecap*

structures capable of withstanding such attack.

Responding to increased U.S. Army emphasis on Arctic defenses, during and after World War II, the U.S. Army Corps of Engineers established laboratories at Wilmette, Illinois, and Boston, Massachusetts, to study the impact of cold climates on military operations. These Corps laboratories conducted research and experimentation on materials and techniques suitable for construction in areas of snow, ice, and permafrost. Their efforts aided the development of the Distant Early Warning (DEW) Line Radar System that stretched across Greenland, northern Canada, and Alaska, as well as the construction of American airfields and bases in those regions. The laboratories consolidated in 1961 to form the Cold Regions Research and Engineering Laboratory at Hanover, New Hampshire.

## U.S. Army Reorganization

In 1962 seeking to streamline the U.S. Army's structure, Secretary of Defense Robert McNamara implemented the most substantial reorganization of the Army in the post-World War II era. The positions of all of the technical service chiefs, except for the Chief of Engineers and the Surgeon General, were abolished, and three newly created functional commands took important responsibilities from the Chief of Engineers. The Army Combat Developments Command assumed responsibility for engineer training and military doctrine. The Office of Personnel

Operations took over the career management of engineer officers and the Army Materiel Command assumed engineer supply and equipment development functions.

Overseeing the development, purchase, and supply of a wide range of U.S. Army weapons and equipment, the Army Materiel Command created a number of major subordinate commands to which it assigned responsibility for specific types of items. The Army Mobility Command (1962–1967) and its successor, the Army Mobility Equipment Command, took over the supply of most military engineering equipment and the supervision of the Engineering Research and Development Laboratories at Fort Belvoir, which became the Army Mobility Equipment Research and Development Center. The two commanders of the Army Mobility Command, Major Generals Alden Sibley and William Lapsley, were both engineer officers, and Sibley moved to the Mobility Command directly from his duties as the last Deputy Chief of Engineers for Military Operations. This eased the transition in engineer supply matters.

Major General William Gribble, later Chief of Engineers, served as the Army Materiel Command's Director of Research and Development in 1964–1966, and Major General Richard Free, another engineer officer, held that position from 1967–1969. These were important years for the development of new engineer materiel used to support American forces in Vietnam. Aided by renewed experimentation in airfield mats and membranes at the Waterways Experiment Station, the Materiel Command developed the prefabricated neoprene-coated nylon membrane, known as the T-17 membrane, used on airfields in Vietnam; new aluminum and steel landing mats; and peneprime, a high-penetration asphalt that met dust-control needs in Vietnam. The Chief of Engineers remained the senior engineer advisor to the Army Chief of Staff; his advice was sought and implemented on such decisions as the selection of the D-7 dozer as the standard bulldozer in Vietnam rather than the newer but less easily transported D-8 model.

Despite its loss of important training, personnel, and materiel supply responsibilities in 1962, the Office of the Chief of Engineers continued to supervise the engineering, construction, and real estate services required by the U.S. Army, U.S. Air Force, and National Aeronautics and Space Administration. The Chief's office also continued to formulate policies governing the maintenance and repair of U.S. Army housing and other real property and the operation of the utilities on Army installations, as it had since World War II. U.S. Army facilities engineers implemented

*Family housing constructed at Ben Guerin, Morocco, in the late 1950s as the Corps built airfields to allow Air Force heavy bombers to reach the Soviet Union.*

these policies under the supervision of installation commanders. The Chief of Engineers, however, lost control of funding in the repairs and utilities sphere in 1958. The Chief of Engineers' work in all of these fields remained under the general staff supervision of the Deputy Chief of Staff for Logistics, while the Assistant Secretary of the Army for Installations and Logistics in 1961 assumed civilian oversight of all of these functions.

In addition, the Office of the Chief of Engineers continued to supervise U.S. Army mapping, geodesy, and military geographic intelligence services, maintaining the Defense Department's worldwide map library, as it had since 1939. Beginning in 1963 and 1964, the office exercised its topographic responsibilities under the program direction of the Assistant Secretary of the Army for Research and Development, with policy guidance from the Army's Assistant Chief of Staff for Intelligence.

While the Engineer Research and Development Laboratories were placed under the Army Materiel Command in 1962, its former topographic and nuclear power development functions remained the responsibility of the Corps of Engineers. With the field of military mapping research expanding rapidly at the dawn of the satellite era, the Chief of Engineers in 1960 transferred this function from the Engineer Research and Development Laboratories to the newly created Engineer Geodesy, Intelligence, and Mapping Research and Development Agency. The reorganization of 1962 left the military mapping agency part of the U.S. Army Corps of Engineers. The

*Autonomous land vehicle, a test robotic vehicle developed by the Engineer Topographic Laboratories, now the Topographic Engineering Center*

agency was renamed the Engineer Topographic Laboratories in 1967.

The Department of Defense consolidated the topographic work of the different military services in 1972, however, and the U.S. Army Topographic Command, whose director had reported to the Chief of Engineers, was absorbed into the new Defense Mapping Agency. The Chief of Engineers again retained responsibility for U.S. Army topographic research and development. The Engineer Topographic Laboratories, located at Fort Belvoir, Virginia, developed during the 1960s and 1970s automated equipment for producing topographic maps from aerial photographs and improved systems of Army field map production. In the 1980s, they developed systems to convert terrain data into digital form and used computer graphics to offer commanders access to this data in a variety of easily interpreted formats. The Corps renamed the Engineer Topographic Laboratories the Topographic Engineering Center in 1991.

The Army Engineer Reactors Group, renamed in 1971 the Army Engineer Power Group, retained the Corps' responsibility for U.S. Army nuclear power development after the 1962 reorganization. In May 1962, the Corps created the Army Engineer Nuclear Cratering Group at Livermore, California, to study, in cooperation with the Atomic Energy Commission, the feasibility of nuclear methods of excavation. Although officials considered using nuclear devices in the construction of a proposed sea-level canal across Central America and in several civil works projects in the United States, no feasible use of this concept was found. The Corps disbanded the Nuclear Cratering Group in 1971.

The Cold Regions Research and Engineering Laboratory was transferred to the Army Materiel Command in 1962, but because of continuing Corps of Engineers requirements for Arctic construction research, the Materiel Command approved its return to the Corps of Engineers in 1969.

After the transfer of the Engineer Research and Development Laboratories to the Army Materiel Command, the Chief of Engineers sought to create a new facility to conduct basic research into construction materials and design, housing habitability and maintenance, and energy and utility systems. As the Ohio River Division's Construction Engineering Laboratory at Cincinnati had begun significant work in this sphere, the Corps, with the approval of the U.S. Army Secretariat, expanded that facility into a new Construction Engineering Research Laboratory. The new laboratory opened in Cincinnati in

1968 and moved the following year to its present location at Champaign, Illinois, where it occupies facilities leased from the University of Illinois. This newest Corps laboratory developed a fibrous reinforced concrete used both in airfield runways and in some civil works projects, a portable instrument to test welding quality, and a centralized facility to control pollutants where U.S. Army vehicles are washed.

In order to streamline its business practices and provide better service to its customers, many of whom were outside organizations, the Corps of Engineers reorganized its research and development laboratories

*Biaxial shock test machine designed by the Construction Engineering Research Laboratory to test both horizontal and vertical structural strength.*

*Construction of "the Pier" in the late 1970s at the Field Research Facility, located near Duck on the Outer Banks, N.C. The facility was part of the Coastal Engineering Research Center, established in 1963 to study coastal engineering problems. After several moves and reorganizations, the center became part of the Coastal and Hydraulics Laboratory of the Engineer Research and Development Center.*

A USACE officer at ERDC using a battlefield computer simulation

into the U.S. Army Engineer Research and Development Center (ERDC) in 1999. The seven component laboratories in ERDC were the Coastal and Hydraulics, Environmental, Geotechnical and Structures, and Information Technology laboratories in Vicksburg, Mississippi (formerly parts of the Waterways Experiment Station); the Construction Engineering Research Laboratory in Champaign, Illinois; the Cold Regions Research and Engineering Laboratory in Hanover, New Hampshire; and the Topographic Engineering Center in Alexandria, Virginia. In the summer of 2006 the Corps continued this process of streamlining and consolidating by combining the positions of Director of the Engineer Research and Development Center and Director of Research and Development in engineer headquarters.

A researcher at ERDC's Environmental Laboratory carrying out experiments in environmental chemistry

## Engineer Troop Units

After World War II, U.S. Army engineer troops were organized primarily into engineer combat and construction battalions, supplemented by topographic battalions and various specialized engineer companies. The combat battalions were designed to provide the engineering capabilities required by front-line forces, and their men were trained and equipped to fight as infantry if necessary. Engineer construction battalions had heavier equipment suited for the more permanent construction typically required to the rear of combat zones, and their members were not expected to fight as infantry. Lieutenant General Walter Wilson, the Chief of Engineers, proposed in 1962 to eliminate the engineer construction battalion and create a single, standardized engineer combat battalion that could be aided, when required for heavier work, by a construction equipment company. The Combat Developments Command studied Wilson's proposal but concluded that the construction battalion would be essential in the event of a lengthy war. Subsequent events in Vietnam supported this conclusion, for engineer construction battalions there played a leading role in building U.S. Army installations and an ambitious highway development program.

The Chief of Engineers regained staff responsibility for the development of Army engineer units in 1969, and a reevaluation of the proper role of the engineer construction battalion soon ensued. The Engineer Strategic Studies Group, a broadly chartered studies and analysis activity reporting to the Chief of Engineers, proposed in 1974 that the engineer construction battalion be reorganized and its firepower augmented so that it, too, would be prepared to assume a full combat role. In the contemporary climate of congressional concern over the military's proportion of combat and support forces, frequently termed the "tooth-to-tail ratio," the U.S. Army then accepted this proposal. Engineer construction battalions at home and abroad were reorganized in 1975 as engineer combat (heavy) battalions. As part of the reorganization, the units were provided additional antitank weapons, grenade launchers, radios, and demolition equipment, and their men were given additional combat training. The conversion of the engineer construction battalions in Europe contributed significantly to the reduction of the U.S. Army's support forces there, as mandated by the Defense Appropriation Act for 1975. In that same year, the U.S. Army again included the U.S. Army Corps of Engineers among its combat arms branches, while also retaining it among its combat support arms and its services.

## Army Facilities Programs

The U.S. Army Corps of Engineers substantially increased its responsibility over the U.S. Army's military construction and family housing programs in 1974. Prior to that time, the Deputy Chief of Staff for Logistics formulated Army budget planning and set basic policies for these facilities programs, which the Corps then executed. The Deputy Chief of Staff for Logistics exercised these functions through his director of installations, as he and his predecessors had done since 1954. As part of a larger transfer of Army staff responsibilities to operating elements, the U.S. Army in 1974 placed the director of installations, Major General Kenneth Cooper, together with his staff and his program development responsibilities, under the Chief of Engineers. General Cooper became Assistant Chief of Engineers. In the same year, the Corps added facilities engineering technical assistance and fossil-fuel energy consulting to the then-dwindling responsibilities of the Army Engineer Power Group, which it renamed the Facilities Engineering Support Agency.

## Environmental Responsibilities

In 1966 the U.S. Army Chief of Staff assigned the Chief of Engineers supervision over the engineering aspects of the Army's emerging program to protect the environment and abate pollution in the construction and operation of its military facilities. He also instructed the Surgeon General and the Chief of Engineers to work together to develop pollution abatement programs for the U.S. Army. In 1971, the deputy chief of staff for logistics assumed primary staff responsibility for directing the Army's environmental preservation and improvement activities, exclusive of the civil works arena. His director of installations created an Environmental Office in that year to undertake this responsibility. The Chief of Engineers continued to supervise the engineering portion of the program.

When the director of installations became the Assistant Chief of Engineers in 1974, the Corps added the direction of U.S. Army environmental efforts related to military sites to those involving civil works projects. This mission came to include supervising the Army's water pollution abatement and solid waste management programs; issuing policies for monitoring and controlling air pollutants emitted by Army facilities and vehicles; and drafting regulations to govern the Army's management of hazardous and toxic materials, its noise abatement efforts, and its responses to any Army-caused oil spills. The Corps also assumed responsibility in 1974 for a U.S. Army program to preserve buildings of historic or architectural significance and noteworthy archaeological sites on Army properties. The Office of the Assistant Secretary of

*The volatilization system that removed dangerous organic compounds that contaminated ground water around the Twin Cities Ammunition Plant, Minn.*

the Army for Civil Works assumed civilian direction of the Army's military environmental program upon the office's establishment in 1975. The Army shifted this oversight function to the office of the Assistant Secretary of the Army responsible for installations and logistics in 1978.

The creation of the Defense Environmental Restoration Program, first funded by a 1983 law, led to a noteworthy enlargement of the Corps' environmental work relating to military installations. The military services had earlier initiated efforts to remove hazardous materials from their active installations. The new program added hazardous waste disposal from former military sites and the removal of unsafe buildings, ordnance, and other debris from both active and former military sites. The U.S. Army Corps of Engineers, which had already begun providing engineering assistance to the Environmental Protection Agency in

*Officers Club at Fort Totten, N.Y., built in the shape of the Corps Castle and placed on the National Register of Historic Places*

*Geo membrane and gas vent layers covering a landfill at K. I. Sawyer Air Force Base, Mich., 1996*

*Removal of contaminated soil, Rocky Mountain Arsenal, near Denver, Colo.*

its direction of civilian toxic waste removal under the Superfund Program enacted in 1980, assumed program management in 1984 of the environmental restoration program for all former military sites, for all services. The deputy for Environmental Policy in the Office of the Deputy Assistant Secretary of Defense for Installations selected sites for cleanup after considering the recommendations of the Office of the Chief of Engineers. This position was raised to Deputy Assistant Secretary of Defense, Environment in 1986.

The U.S. Army Toxic and Hazardous Materials Agency, created in 1978 at Aberdeen, Maryland, as a subordinate activity of the Army Materiel Command, maintained operational control of the expanded environmental restoration program on active U.S. Army installations. It also relied on the U.S. Army Corps of Engineers for most of its design and construction work. The Corps had provided similar assistance in the cleanup of many active U.S. Air Force installations. In 1988, the Army placed the Toxic and Hazardous Materials Agency under the Chief of Engineers, consolidating Army environmental responsibilities under a single head.

*Demolition of a smokestack at a Bunker Hill, Idaho, Superfund site, 1995*

## Army Facilities Maintenance

The U.S. Army Corps of Engineers increased its involvement in maintaining and repairing Army housing and other facilities at the same time it broadened its environmental responsibilities. A study panel headed by engineer Lieutenant General Lawrence Lincoln in 1968 urged the U.S. Army to encourage installation facilities engineers to turn to Corps districts and divisions for engineering support by funding a portion of that work. The U.S. Army agreed to set aside a modest fund for Corps installation support, invited installation commanders to turn to the Corps for additional maintenance and repair work on a reimbursable basis, and took other actions recommended by the Lincoln Panel to strengthen facilities engineering.

When the administration of President Jimmy Carter proposed management consolidation and increased reliance on private-sector contracting in the maintenance of U.S. Army facilities, the U.S. Army Corps of Engineers undertook several new studies in this sphere. A panel headed by Brigadier General Donald Weinert reviewed Army facilities engineering in the context of the era's heightened emphasis on master planning, energy conservation, worker safety, and environmental protection. The group observed in 1978 that the Corps' resources were still often neglected in the facilities maintenance sphere, despite the U.S. Army's implementation of most of the Lincoln Panel's recommendations. A subsequent engineer planning group headed by Colonel Charles Blalock proposed incorporating installation facilities engineers into the Corps' district organization, aiding them with the Corps' substantial experience in contracting, and giving them a full range of local engineering responsibilities.

Although the U.S. Army did not accept the offer of Lieutenant General John W. Morris, Chief of Engineers, to assume such broad installation engineering responsibilities, it did approve the plan, elaborated by the Engineer Studies Center (formerly the Engineer Strategic Studies Group), to centralize Army facilities maintenance work in the Military District of Washington under a single engineer manager. The U.S. Army Corps of Engineers in 1980 created the Engineer Activity, Capital Area, at Fort Myer, Virginia, to exercise that function.

Although installation commanders retained responsibility for maintenance work on U.S. Army posts, their facilities engineers turned increasingly to Corps districts and divisions for assistance in prosecuting the Reagan administration's substantial effort to reduce the backlog of Army repair and maintenance work. Streamlining its procedures in this sphere, the U.S. Army Corps of Engineers saw its reimbursable installation support work grow from $130 million in 1980 to $620 million in 1986. Effective Corps support in this work was enhanced by new administrative reforms proposed by internal reviews made in 1985 and 1988, the former by a panel headed by North Central Division Engineer Brigadier General Jerome Hilmes, and the latter by the Office of the Engineer Inspector General, Colonel Dennis Bulger.

## A Major Command

Witnessing a decline in support for large, new water resources projects in the later 1970s, Chief of Engineers Morris attempted to strengthen his office's ties to the U.S. Army as a whole. Consequently, in 1979 the U.S. Army Corps of Engineers—comprising the Office of the Chief of Engineers and the divisions, districts, laboratories, and other agencies subordinate to the Chief of Engineers—was designated an Army major command. This status gave the Corps a position comparable to other leading specialized Army commands such as the Training and Doctrine Command, Materiel Command, Communications Command, and Health Services Command, and the Army components of unified commands, such as U.S. Army, Europe, and the Eighth Army in South Korea.

The Chief of Engineers' ties to the U.S. Army were strengthened further in 1986 when he was named Chief of the Corps of Engineers Regiment, a ceremonial institution through which all engineer Soldiers, officers, and units would participate in the new U.S. Army Regimental System. The Chief of Engineers' assumption of this position gave symbolic recognition to his office's long history of leadership among the U.S. Army's military engineers.

The Goldwater-Nichols Department of Defense Reorganization Act of 1986 obliged the U.S. Army to distinguish clearly between the small group of personnel who continued to serve the Chief of Engineers in his capacity as an Army staff officer, and the larger number who worked for him as commander of the U.S. Army Corps of Engineers, the engineering and construction organization. The act also mandated personnel reductions that had an impact on the Office of the Chief of Engineers as an Army staff office. Responding to both the Army staff personnel limitations and his own view of current management requirements, the Chief of Engineers, Lieutenant General E. R. Heiberg III, ordered the consolidation of the Facilities Engineering Support Agency and the technical support activities of the Assistant

*Distinctive Unit Insignia of the U.S. Army Corps of Engineers as a major Army command*

Chief of Engineers in the fields of facilities engineering and housing management. The new organization resulting from the consolidation, called the U.S. Army Engineering and Housing Support Center, was established in 1987 at Fort Belvoir, Virginia. Its creation left U.S. Army program development responsibilities in the facilities and housing spheres in a leaner Office of the Assistant Chief of Engineers, now distinctly an Army staff organization. The Army Environmental Office became an Army staff support agency, which also reported to the Assistant Chief of Engineers. The new Engineering and Housing Support Center assumed responsibility for providing engineering support and technical policy interpretation for facilities and housing to U.S. Army forces worldwide.

*Family housing, Fort Belvoir, Va.*

*Dormitory, Fort Huachuca, Ariz.*

In addition to supporting U.S. Army installations at home and abroad, the Corps undertook a major new responsibility for supporting the Army with facilities and services during military operations. After the Cold War ended and the U.S. Army demonstrated its clear military superiority on the conventional battlefield during the Gulf War of 1990–1991, it was not clear what military challenges the new era would bring. However, with pressure to reduce the size of the military, the U.S. Army's leaders emphasized moving uniformed personnel to combat positions and relying on civilian contractors to perform more support services.

The U.S. Army Corps of Engineers, in cooperation with the Department of the Army's Deputy Chief of Staff for Logistics, developed a contract that would use a civilian contractor to prepare plans and perform selected services to augment

U.S. forces during military contingency operations overseas. Based on the Army's newly created Logistics Civil Augmentation Program (LOGCAP), which had been conceived in the 1980s, the contract was broadly structured to cover a number of scenarios worldwide requiring varying levels of support to U.S. military forces based on the theater commander's needs. The Army set up the contract to provide basic life support, maintenance, and transportation services. The Corps' Transatlantic Division awarded the first LOGCAP contract (LOGCAP I) in August 1992, and it was used to support U.S. and United Nations forces sent to Somalia in December 1992.

In total, U.S. forces used LOGCAP I to support six contingency operations from 1992 through 1997, including the largest operation, which was in Bosnia. In 1995 North Atlantic Treaty Organization forces, including American troops, entered Bosnia on a peacekeeping mission. LOGCAP I was used in the Balkans from December 1995 through May 1997.

During this time, the U.S. Army transferred official responsibility for LOGCAP program management to the Army Materiel Command, effective October 1996. Because the U.S. Army Corps of Engineers remained responsible for the first five-year contract, and since the peacekeeping operations had been extended in Bosnia, U.S. Army Europe (USAREUR) asked the Transatlantic Division, now known as the Transatlantic Programs Center (TAC), to award a follow-on logistics services contract. From May 1997 through May 1999, logistics services were provided under a sole source contract to avoid any disruption of services to U.S. forces in the Balkans.

With the commitment of U.S. forces for an indefinite period, USAREUR asked TAC to competitively award the Balkans Support Contract with a contract period of May 1999 through May 2004. Meanwhile U.S. troops entered Kosovo in 1999, and the new Balkans Support Contract, which was separate from LOGCAP, provided logistics support services for operations in both Bosnia and Kosovo. Subsequently, the Balkans Support Contract was extended to accommodate a protracted evaluation period. Ultimately TAC awarded the follow-on Balkans Support Contract in June 2005. While the Corps continued to support USAREUR with managing its logistical services contract requirements, USACE did not have official responsibility for LOGCAP after the Army transferred the program to the Army Materiel Command in 1996.

## Restructuring of Installation Support

As the U.S. Army turned more of its attention to its domestic installations in the aftermath of the Cold War, Acting Secretary of the Army John Shannon in 1993 gave broad authority over planning, programming, and general support for Army bases, facilities, and environmental restoration efforts to a new assistant chief of staff for installation management. This new Army staff officer assumed most of the responsibilities of the Assistant Chief of Engineers, whose office was abolished. The Army Environmental Office, the Army Environmental Center (as the U.S. Army Toxic and Hazardous Materials Agency had been renamed), and elements of the Engineering and Housing Support Center involved in policy were also placed under the new Assistant Chief of Staff. General officers, who had previously reported to the Chief of Engineers, became the first directors of Environmental Programs and of Facilities and Housing for the Assistant Chief of Staff for Installation management. The military engineering and topography functions that had been overseen by the Assistant Chief of Engineers, however, remained Army staff responsibilities of the Chief of Engineers. They were henceforth exercised by the newly established Office of the Chief of Engineers (Pentagon). The Engineering and Housing Support Center was renamed the U.S. Army Center for Public Works. Remaining under the Chief of Engineers, it has continued to provide technical support to installation commanders. Overall, the U.S. Army Corps of Engineers retained its design and construction missions, including the execution of a large and expanding program for the cleanup of hazardous materials at current U.S. Army and U.S. Air Force installations and former defense sites.

In 1998 the headquarters of the U.S. Army Corps of Engineers began its own major restructuring of the installation support mission. The Center of Public Works became the Installation Support Center in preparation for abolishing the organization and establishing two elements in its place. In 1999 the Corps established an Installation Support Division as one of four major divisions in the Directorate of Military Programs. The new division oversaw real property facilities management and installation support activities for the Directorate of Military Programs and provided related services for the Assistant Chief of Staff for Installation management and the U.S. Army. Other members of the Installation Support Center were sent forward to engineer divisions, where they would be located closer to their customers and could provide more effective installation support.

Customer support became even more important in 2002, when the U.S. Army instituted one of the most fundamental changes in the management of installations in its history. In spite of attempts to centralize installation management, including one by the powerful Army Service Forces during World War II, the U.S. Army persisted in the policy of assigning the senior combat commander on an installation the additional duty of installation commander. With the establishment of the Installation Management Agency as a field operating agency of the Assistant Chief of Staff for Installation Management, the Army split the two functions, establishing a separate garrison commander responsible to the Installation Management Agency. The combat unit commander could concentrate on his military mission, leaving the Installation Management Agency responsible for establishing the standards and providing the resources to ensure equitable services and quality of life on all U.S. Army installations. The U.S. Army Corps of Engineers now works closely with the Assistant Chief of Staff for Installation Management and the Installation Management Agency to perform its military construction responsibilities for the U.S. Army, one of the Corps' key missions since the beginning of World War II.

## Corps and Army Restructuring

In 2006 the U.S. Army Corps of Engineers (USACE) was undergoing an organizational transformation from a major U.S. Army command, which it had become in 1979, to a direct reporting unit (DRU). In a major restructuring that went into effect in the summer of 2006, the Army abolished the major Army command (MACOM) as an organizational element and transferred all old MACOMs and several new organizations to one of three categories: Army Commands, Army Service Component Commands, and Direct Report Units (DRUs).

Three former MACOMs—Training and Doctrine Command, Forces Command, and Army Materiel Command—became Army Commands. Nine Army component commands, such as U.S. Army Europe, U.S. Army Pacific, U.S. Army Central, and Eighth U.S. Army, became Army Service Component Commands. Eleven Army organizations, including several of the remaining former MACOMs, such as USACE, and a number of other organizations, such as the Installation Management Agency and the Acquisition Support Center, became DRUs.

DRUs are Army organizations with institutional or operating functions that provide broad general support to the Army, usually in a single, unique discipline. DRUs report to a member of the Army staff, but since the Chief of Engineers was both an Army staff officer and the USACE commander, his status in this regard remained unchanged. USACE's lineage and heraldic honors and insignia also were preserved. An implementing Army general order was expected by the end of 2006.

According to the Army announcement issued on June 6, 2006, the restructuring was intended to contribute to the process of Army transformation and increase the Army's responsiveness at home and abroad. By summer 2006 the Corps of Engineers was undertaking a huge, multi-year military construction and base realignment and closure workload for the Army and the Air Force and providing major support to the effort to rebuild Iraq and Afghanistan. The engineers' domestic and global responsibilities remained large and diverse as it supported the U.S. Army and the Nation.

# Constructing Camp Bondsteel

Following the successful bombing campaign launched by nations of the North Atlantic Treaty Organization to induce Serbia to cease ethnic cleansing operations in Kosovo, during the summer of 1999 U.S. military forces entered the province to provide security and protect Kosovar refugees. Called Task Force Falcon, this force required extensive headquarters, logistical, operational, and housing facilities, which U.S. Army engineers provided.

The commander of the engineer brigade, 1st Infantry Division, Colonel Joseph Schroedel, who later became commander of the South Pacific and South Atlantic divisions, oversaw the initial construction effort to support the deployment of Task Force Falcon. Building the Kosovo base camps involved some 1,700 military engineers augmented by 1,000 employees of Brown and Root Services under a logistics support contract managed by the Corps of Engineers. The Waterways Experiment Station provided data for locating water sources. A team from the Baltimore District advised on environmental engineering and demining. Nearly 7,000 local skilled and unskilled laborers assisted the U.S. Army engineers in base construction.

These engineer troops constructed four base camps in the region and two large ones in Kosovo. The latter were Camp Bondsteel and a smaller base camp nearby. Staff Sergeant James L. Bondsteel received the Medal of Honor during the Vietnam conflict. The majority of the construction at Camp Bondsteel, built from the ground up on a former farm field, occurred in just three months. Con-

*Soldiers of the 320th Engineer Company set up a positioning receiver to survey the airfield, Camp Bondsteel, Kosovo. The low building in the left, center, is a SEA hut.*

*Department of Defense*

struction proceeded twenty-four hours a day during that time.

The U.S. Army decided to utilize rapidly constructed, semipermanent Southeast Asia (SEA) huts to provide troop housing quickly. The SEA huts, which got their name from previous wartime employment in Southeast Asia, were modified for use in the Balkans. Each SEA hut was ninety-two feet long by thirty-two feet wide and included five sleeping rooms plus a combination shower and latrine. The temporary units were made of plywood with metal roofs. Rooms had wall-mounted heating/cooling systems, electricity, and a drywall finish.

*Camp Bondsteel, Kosovo*

Although the engineer brigade returned to the United States in 2000, the support of the U.S. Army Corps of Engineers continued thereafter. Camp Bondsteel, near Urosevac, Kosovo, subsequently served as headquarters for the Multi-National Brigade (East). Over time Camp Bondsteel has evolved into what is by any measure an immense post. Its perimeter measures 7 miles and encloses an area of 955 acres. In the construction of the base, 20 miles of roads were built, 100 miles of electrical cable were laid, and a half-million cubic yards of earth were moved. The post is divided into two sections: North Town and South Town. Approximately 5,000 Soldiers live in more than 250 SEA huts. Also on post are a 30,000-square-foot headquarters building, an ammunition dump, motor pools, chapels, recreation and dining facilities, an operations center, two post exchanges, a wastewater treatment plant, and a heliport. The U.S. Army Corps of Engineers also helped design force protection structures for the base.

*Idealized view of water management*
*President's Water Resource Policy Commission*

# Civil Works, Congress, and the Executive Branch

## Early Civil Works Oversight

From the earliest beginnings of the U.S. Army Corps of Engineers, both Congress and the cabinet official overseeing the U.S. Army carefully monitored and guided the involvement of the Corps in civil works projects. In fact, in 1800, it was Secretary of War James McHenry who suggested that engineer officers possess talents that serve the country not only in war, but also in peacetime "works of a civil nature."

Once the Corps was permanently established in 1802, few operational and organizational changes were made without the explicit authorization of the Secretary of War. Indeed, the Chief of the Engineer Department, along with the chiefs of other War Department bureaus, enjoyed direct access to the Secretary of War and protested vehemently whenever the U.S. Army's commanding general attempted to interfere with that access. Even the correspondence procedures reflected this close relationship. Mail intended for the Chief Engineer was sent under cover to the Secretary of War with the words "Engineer Department" written on the lower left-hand corner of the envelope. Conversely, reports from the U.S. Army engineers intended for Congress were transmitted through the Secretary of War. The precise role of the U.S. Army commanding general was not clarified until Congress abolished that position and created the position of chief of the general staff at the beginning of the twentieth century.

Examples of early oversight activities of the Secretaries of War are numerous. John C. Calhoun did not hesitate giving guidance to the Board for Internal Improvements, organized in 1824 to administer the responsibilities imposed by the General Survey Act. Charles M. Conrad transferred certain civil works responsibilities from the Topographical Engineers to the Corps of Engineers following passage of the 1852 Rivers and Harbors Act. His successor, Jefferson Davis, allowed the use of local funds to continue projects that had already received some congressional appropriations. In these and other ways, the Secretaries of

*James McHenry, Secretary of War, 1796–1800*

*Charles M. Conrad, Secretary of War, 1850–1853*

*U.S. Army Center of Military History*

War profoundly influenced the organization and direction of the U.S. Army engineers.

Meanwhile, Congress also helped mold the operations and policies of the U.S. Army Corps of Engineers. Congress not only appropriated funds and authorized civil works projects, it also specified how many officers the Corps was to have, conditions for their promotion, and even how much per diem (if any) they should earn while assigned to a project. Congress authorized oversight boards of engineer officers and determined what precise responsibilities the boards were to discharge. It requested surveys and reports, and congressional committees carefully reviewed the Corps' progress on its civil works assignments, rarely failing to call attention to a real or imagined defect in the Corps' management. The responsibility of the Engineer Department to carry out the wishes of Congress, including the development of "internal improvements," was explicitly noted in the General Regulations of the Army as published in 1825.

After the Civil War, the congressional role in Corps affairs became even more evident. While not appreciably increasing the number of officers assigned to the Corps, Congress substantially increased the Corps' work on rivers and harbors. Consequently, the Corps was forced to depend on help from the civilian engineer community. This dependence worked to the Corps' disadvantage. Most of the civilian engineers did not become career employees of the Corps, but the very fact of their employment helped give credibility to the charge that the Corps was unable to fulfill its civil works functions. Civilian engineers maintained that they, not military engineers, should be in charge of civil works. They lobbied Congress, and their congressional sympathizers introduced numerous bills beginning in the 1870s to transfer civil works functions from the U.S. Army Corps of Engineers to some other part of government; often, the preferred solution was to create a new Department of Public Works. Railroad interests, which perceived the Corps as an unfair competitor in the development of national transportation systems, wished to have the private sector do all river and harbor work. Pummeled from many quarters, the Corps saw its relationship with Congress become more dependent and more fractious.

Authorizations and appropriations during this period reflected some of the worst evils of pork-barrel legislation. Projects were poorly chosen, piecemeal appropriations were commonplace, and the U.S. Army Corps of Engineers often gave unreliable estimates. Around the turn of the century, relations improved, mainly as a result of the work of Ohio Representative Theodore E. Burton. As chairman of the Rivers and Harbors Committee, he shepherded through Congress a bill establishing the Board of Engineers for Rivers and Harbors within the U.S. Army Corps of Engineers to examine costs, benefits, and necessity of river and harbor improvements. In the 1907 Rivers and Harbors Act, Burton did not allow one new project to be added unless the entire cost of the project was appropriated and it had the express approval of the Chief of Engineers. Burton's efforts briefly curtailed pork-barrel legislation, but when he left the House of Representatives for the Senate in 1909, Congress quickly reverted to its old ways. The 1910 Rivers and Harbors Act appropriated funds for projects in 226 of the 391 congressional districts.

## Secretary of War's Role

While Congress busily gave the Corps work, the Secretaries of War attempted to oversee the Corps' execution of its civil works projects. This attention to Corps operations may have been a matter of choice with some Secretaries, but several rivers and harbors acts passed in the 1880s explicitly mandated that the Secretary of War supervise the expenditure of appropriated funds to, in the words of the 1884 act, "secure a judicious and economical expenditure of said sums." The Secretary was directed furthermore to submit to Congress annual reports of work done, contracts made, and funds expended.

Pursuant to these acts, the Secretary of War issued new regulations in 1887 that specifically delegated to the Chief of Engineers the responsibility to supervise "all disbursements by officers of the Corps." Slightly modified in 1889, these regulations also charged the Chief of Engineers to present to the Secretary of War an annual report of Engineer Department operations and, "with the approbation of the Secretary of War," to determine the quality, number, and physical characteristics of equipment needed by the U.S. Army engineers. The Secretary of War approved the assignment of division engineers as well as officers to serve on the board that oversaw fortifications and river and harbor improvements. He approved the initiation of new projects and specified the forms to be used to contract work. Moreover, he approved any modifications of the original contract. Finally, it should be noted that it was the Secretary of War, not the Chief of Engineers, who Congress charged to have surveys done, civil works projects constructed, and rules issued to regulate federally operated

*Theodore E. Burton, Congressman and Senator from Ohio*

Senate Historical Office

*Francis G. Newlands, Senator from Nevada*

canals and waterways. The work, of course, was then assigned to the U.S. Army Corps of Engineers.

## Multipurpose Water Management

In the Progressive Era at the beginning of the twentieth century, the Secretary of War's office became embroiled in the controversy over the development of multipurpose water projects. Multipurpose planners sought to develop coordinated river basin programs that responded to a wide variety of needs, including navigation, flood control, irrigation, water supply, and hydropower. The U.S. Army Corps of Engineers generally opposed the concept, arguing that other purposes should always be subordinated to navigation in federal projects, that multipurpose dams would be difficult to operate, and that greater coordination was not needed; existing government agencies could provide whatever coordination was required.

However, multipurpose development supporters had powerful friends in Congress, especially Senator Francis G. Newlands of Nevada, who introduced legislation to establish a multipurpose water resources coordinating commission. Henry L. Stimson, President William H. Taft's Secretary of War, was an avid conservationist and a former member of the board of directors of the National Conservation Association. He wholeheartedly supported the Newlands measure. So did Newton D. Baker, who served as Secretary of War under President Woodrow Wilson. Other Secretaries, such as Taft himself, who headed the War Department before he succeeded Theodore Roosevelt as president, and Lindley M. Garrison, who served in Wilson's first administration, were more sympathetic toward the Corps.

Secretary of War Stimson complained about his relationship with the Chief of Engineers. Stimson asked the Chief whether an improvement should be made in light of other demands on the budget. Without answering the question, the Chief of Engineers, Brigadier General William H. Bixby, simply responded that the project was good for the country without comparing it with other projects or budgetary demands. Stimson pursued his point. He wanted to use a comparative approach. However, Bixby objected, "I have nothing to do with that. I cannot have anything to do with it. Congress will not listen to me on that. They reserve the judgment to do that themselves." Stimson thought the Corps was uncooperative and unresponsive, but there was some merit in the argument of the Chief of Engineers.

As Newlands himself pointed out, numerous rivers and harbors acts had indeed constrained the Corps' flexibility. Although the Corps had authority only to recommend a project based on its own merits, it did seem to support projects that were politically feasible and not necessarily urgently required. Also, the Corps' opposition to a more constructive, integrated approach to water resources management reflected a predictable bureaucratic concern for maintaining maximum administrative independence. Despite some initial legislative success, Newlands saw his plans for a great waterways commission unravel when the U.S. declared war on Germany in April 1917.

The 1925 Rivers and Harbors Act accelerated the movement toward multipurpose water management. It authorized the Corps and the Federal Power Commission to prepare cost estimates for surveys of navigable streams and tributaries "whereon power development appears feasible and practicable." The aim was to develop plans to improve stream navigation "in combination with the most efficient development of the potential water power, the control of floods, and the needs of irrigation." The Corps responded with a recommendation for 24 surveys at an estimated cost of $7.3 million.

In 1927 Congress appropriated the necessary funds, whereupon the Corps launched a series of comprehensive river surveys. The resulting reports, known as the "308 Reports" after the House document in which the survey estimates had first appeared, became basic planning documents for many of the multipurpose projects later undertaken by the federal government. During the depths of the Great Depression, Congress authorized the Corps to supplement the 308 Reports with studies "to take into account important changes in economic factors as they occur and additional streamflow records or other factual data." This authority charged the Corps with a broad responsibility to undertake continuing river basin planning, with an emphasis on navigation and flood control.

## Relationship with Congress

From about 1885 to 1925, Americans' daily lives were more and more affected by the federal government. Working with the executive branch, Congress attempted to control abuses that could threaten the liberty, livelihood, or health of the citizenry. To do so, it was necessary to increase the regulatory authority of various federal agencies, including the War Department. In 1886, Congress gave the Secretary authority to regulate harbor lines. The 1890 Rivers and Harbors Act expanded the Secretary's authority to regulate and remove any

*Henry L. Stimson, Secretary of War, 1911–1913; 1940–1945 and Secretary of State, 1929–1933*
U.S. Army Center of Military History

George H. Dern, Secretary of War, 1933–1936
U.S. Army Center of Military History

navigation obstructions, including bridges, waste material, and structures such as dams and piers built outside of established harbor lines. In 1894, Congress authorized the War Department to regulate navigation in all federally owned canals, whether or not the Corps had built them. The 1899 Rivers and Harbors Act gave the Secretary added authority to regulate the dumping of waste material into navigable streams and the construction of any structures that might impede navigation. The 1906 General Dam Act authorized the Secretary of War to review and approve plans and specifications for all dams to be constructed across navigable waters. While, of course, most of these new responsibilities were delegated to the Corps of Engineers, in no case did Congress bypass the Secretary and grant power directly to the Chief of Engineers.

The Corps' relationship with Congress in the interwar period was extremely close. Indeed, Secretary of War George H. Dern called the Corps "an agency of the legislative branch" in a 1934 report to the president. Congress did not just establish overall water resources policy, but congressional committees also determined which projects should be funded and the extent and timing of the funding. One procedure that was used extensively was the committee review resolution, which required the Corps to reconsider reports in which it had recommended against project construction. This was a particularly popular device during the New Deal, when projects were needed for work relief as well as for navigation or flood control. For instance, only about one-third of the projects authorized in the 1935 Rivers and Harbors Act originated as favorable reports. Reports on most of the others had been modified in response to a committee review resolution. The procedure constituted a kind of quasi-legislative process that circumvented both the rest of Congress and the executive branch.

Corps orders and regulations directed district engineers to contact each member of Congress within their districts to solicit the congressman's wishes about river and harbor improvements. The congressman was also invited to testify at a public hearing dealing with the project and to present written arguments to the Board of Engineers for Rivers and Harbors, which reviewed the project report. If the congressman was still dissatisfied, then he always had recourse to the committee review resolution. Although this kind of relationship could have led to tension, such was not the case. Congressmen protected the Corps at the same time they pressured it. All efforts by President Franklin D. Roosevelt to centralize water resources planning and institute some Progressive Era ideas met immovable congressional (and War Department) opposition; the Corps remained the water resources agency of choice in both wings of the Capitol.

The passage of the 1936 Flood Control Act, which officially recognizing a federal obligation in flood control activity, greatly expanded the responsibilities of the U.S. Army Corps of Engineers. The law authorized the expenditure of $320 million for about 250 projects and a number of examinations and surveys. Since 1936, the Corps has built, pursuant to congressional authorizations and appropriations, more than 300 reservoirs whose primary benefit is flood control.

## Policy Coordination Efforts

More so than any of his predecessors, President Roosevelt attempted to ensure interagency coordination of federal water projects. In 1939, he instructed the departments of War, Interior, and Agriculture to cooperate with his National Resources Planning Board in drawing up a memorandum that would ensure consultation among all federal water agencies during project planning. The subsequent tripartite agreement resulted in a better and more efficient exchange of information among the agencies; however, it failed to eliminate bureaucratic rivalries.

Roosevelt finally gave up on developing a centralized natural resources planning organization in 1943 when Congress refused to appropriate money to keep the National Resources Planning Board in existence. However, the president continued to press one of the board's chief ideas, basinwide planning commissions such as the Tennessee Valley Authority established ten years earlier. His support of the

*President Franklin D. Roosevelt*
*National Archives*

*Harry S. Truman*

Missouri Valley Authority reflected this commitment. A similar authority for the Columbia River Basin was discussed, and Roosevelt's successor, Harry S. Truman, embraced the idea. Nevertheless, continued congressional skepticism assured that river basin commissions would never obtain the authority that Roosevelt and Truman envisioned.

Although Congress effectively destroyed the National Resources Planning Board during the war, federal agencies continued to coordinate their various responsibilities. The Departments of War, Agriculture, and Interior established the Federal Interagency River Basin Committee (FIARBC), commonly called "Firebrick." Later, the Departments of Labor and Commerce and the Federal Security Agency (which supervised the U.S. Public Health Service) joined. Various technical subcommittees attempted to coordinate water development in specific river basins, usually meeting limited success. In 1954, President Eisenhower replaced the commission with the new Interagency Committee on Water Resources (IACWR). "Icewater," as this agency became known, had minimal impact because its objective of strengthening executive authority elicited little interest in Congress.

The various official committees and study commissions, like the first and second Hoover Commissions of the post-World War II period, mirrored an emerging consensus that rational water resources development required uniform procedures and ongoing coordination. However, executive branch committees such as Firebrick lacked the clout to be effective interagency vehicles. The organization in the executive branch that did seem to have the necessary visibility and bureaucratic authority was the Bureau of the Budget, later renamed the Office of Management and Budget. Upon the dissolution of the National Resources Planning Board in 1943, President Roosevelt issued Executive Order 9384, which directed all federal public works agencies to submit their updated long-range programs directly to the Bureau of the Budget. The major goal seemed to be to ensure that the bureau had the opportunity to see how well agency long-range plans fit into the overall administration program. Although the budget bureau attempted to create a new division to handle the review of agency programs, Congress refused to appropriate funds to hire personnel. Therefore, the bureau was forced to review the programs with existing personnel, and the result was a limited review that ignored such issues as the conformance of agency water project plans with regional plans, social utility, or reliability of the cost/benefit analysis.

Nevertheless, in December 1952, the Bureau of the Budget drafted a far-reaching directive pertaining to the planning of water projects. Known simply as Circular A-47, the document stipulated that the benefits of each element in a multipurpose project must exceed the costs; it would no longer suffice for the total benefits to exceed total costs. Circular A-47 also directed that 50 years would be the maximum allowable time for the repayment of a federal investment. Although criticized in Congress, the guidance remained the basic planning document for the next decade and placed the Bureau of the Budget in the middle of the ongoing debate over water resources planning.

The Eisenhower administration attempted to place individual projects in the context of other national priorities and was skeptical of massive dam-building projects. The Bureau of the Budget generally looked far more favorably at smaller urban flood control projects. Moreover, budget personnel advocated reducing the planning period, if at all possible, to move ahead with actual construction. Of course, Congress could and often did insert projects into bills that not only had not received bureau approval, but had not even been recommended by the Corps of Engineers. For instance, a 1956 bill vetoed by Eisenhower would have authorized thirty-two projects that had not been reviewed by the Corps. A 1958 bill, also vetoed, would have authorized four projects, costing $27 million, that had no project reports, and another three projects, costing $115 million, that had a negative cost/benefit ratio. In 1959, Congress passed a bill over a presidential veto. Eisenhower had disapproved the bill because of the expense involved, some $800 million.

*Dwight D. Eisenhower*

Budgeting (PPB) System throughout the federal agencies. First advanced by Secretary of Defense Robert McNamara in the Pentagon, the program was designed to allow for closer oversight of executive programs.

Although few federal agencies reacted enthusiastically to the presidential order, one that did was the Army's Office of Civil Functions. In 1965, Fitt established a Systems Analysis Group to develop new procedures for preparing the civil works budget and to draft a long-range water investment program for the Nation. Group members proposed to shift emphasis from individual projects—the details of which were familiar only to the members of Congress directly concerned—to water resources problems in the various regions of the Nation. Under Robert E. Jordan III, U.S. Army general counsel and special assistant to the Assistant Secretary for Civil Functions, the Systems Analysis Group perfected a budgeting system and a five-year investment program based on regional allocations. This new approach was firmly installed in the Corps.

Ultimately, however, neither the Bureau of the Budget nor Congress proved capable of shedding the project-by-project orientation in favor of a more programmatic approach to civil works budgeting. Still, the creation by Fitt and the use by Jordan of the Systems Analysis Group initiated an oversight and broadening of the Corps' civil works program that was far removed from the benign neglect of the preceding decade, and it presaged the establishment of the position of Assistant Secretary for Civil Works.

Utah Senator Frank E. Moss's attempt to establish a Department of Natural Resources, which would have included the Corps' civil works functions, and the nearly successful attempt in 1968 to put a congressional moratorium on public works projects signified the gradual dissolution of the Corps' traditionally strong water resources constituency in Congress. Under Jordan, and with the powerful support of Jordan's capable successor, Under Secretary of the Army Thaddeus Beal, the Systems Analysis Group pressed for new Corps missions: wastewater management and urban studies. Although these initiatives failed to produce new construction responsibilities for the Corps, the experience showed that a secretarial-level political appointee, who focused on civil works, would be of enormous benefit. That appointee could help strengthen planning and review functions within the Corps, and concurrently, give the Corps more clout within the executive branch, such as in the interdepartmental Water Resources Council, established in 1965.

Mainly through the efforts of California Representative Don

Clausen, Congress inserted a section in the 1970 Flood Control Act that authorized the position of Assistant Secretary of the Army, Civil Works; however, it was to be another five years before the executive branch appointed the first Assistant Secretary. This was largely because President Richard Nixon supported the creation of a new Department of Environment and Natural Resources and did not wish to do anything that appeared to strengthen the Corps' civil works mission. Finally, on March 20, 1975, Victor V. Veysey, a former representative from California, was sworn in as the first Assistant Secretary of the Army for Civil Works. He served until January 1977.

Victor V. Veysey
Department of Defense

## Role of the Assistant Secretaries

Veysey had the difficult task of defining both his mission and his relationship with the U.S. Army Corps of Engineers. His approach was to act the "honest broker" between the Corps and other organizations involved with water resources; it was an approach that succeeding Secretaries emulated. While working as a conduit between the Corps and its environmental opponents, Veysey never lost the high respect he held for the Corps. He acted forcefully on certain issues, but he looked upon his role primarily as an advisory one. "I wasn't about to order the Chief of Engineers to do anything because I couldn't; that wasn't my role. He takes his orders from the Army chief of staff. But influence, yes. We could try to influence him in directions and in policy, procedure, and so forth.... But from the post of Assistant Secretary you don't order the Chief of Engineers to do anything."

President Jimmy Carter, who questioned the necessity of many water projects and emphasized environmental concerns, did not appoint an Assistant Secretary until April 1978. He chose Michael Blumenfeld, who also served as Deputy Under Secretary of the Army. The Senate failed to confirm Blumenfeld as Assistant Secretary until April 1979. Working through the Water Resources Council, he exerted strong leadership to develop new, environmentally

Michael Blumenfeld
Department of Defense

*William R. Gianelli*
*Department of Defense*

*Robert K. Dawson*
*Department of Defense*

*Robert W. Page*
*Department of Defense*

sensitive principles and standards to guide the planning of water projects.

With the transfer of power from a Democratic to a Republican administration in 1981 came new water resources priorities. The new Assistant Secretary for Civil Works, William R. Gianelli, had formerly headed California's Department of Water Resources under then-Governor Ronald Reagan. His objectives were to reform the regulatory program and to develop new ways to fund the Corps' water resources projects. Both objectives reflected political and philosophical shifts. Gianelli considered the Corps' responsibility to regulate the dredging and filling of wetlands a water quality issue and not a mandate to protect wetlands. He changed regulatory procedures to shorten the processing time, partly by limiting the traditional way of appealing permit decisions. He also led early Reagan administration efforts to reduce the federal financial burden in activities that he believed nonfederal interests could and should fund.

Gianelli's work, together with an unexpected positive response by project sponsors, helped convince Congress that some sort of cost-sharing was necessary if sound water projects were to proceed. It fell to Gianelli's successor, Robert K. Dawson (appointed Acting Assistant Secretary in May 1985), working with Congress, to bring the process to a successful conclusion. The Water Resources Development Act of 1986, signed into law on November 17, 1986, signaled a major historical change in the financing of water projects by requiring cost-sharing for most projects. At the same time, the act authorized about 300 new water projects and numerous studies at an estimated cost of more than $15 billion.

Under Dawson's successor, Robert W. Page, the Corps addressed a wide range of subjects to make project development—from planning through construction—more efficient, faster, and cheaper, without sacrificing quality. The Corps rewrote planning procedures to ensure that nonfederal project sponsors, principally states and local communities, were full partners in project development. After Page left office in October 1990, the position remained vacant until July 1991, when Nancy Dorn became the first female Assistant Secretary of the Army for Civil Works. Perhaps more than her predecessors, Dorn was conservative about seeking new missions. She emphasized instead effective management of the Corps' existing missions during her tenure, which lasted until January 1993.

Under Assistant Secretaries Dorn and Page, the Corps undertook major reforms of the wetlands regulatory program. Policy guidance and

*Nancy P. Dorn*

*Department of Defense*

*President Ronald Reagan signs the Water Resources Development Act of 1986.*

*National Archives*

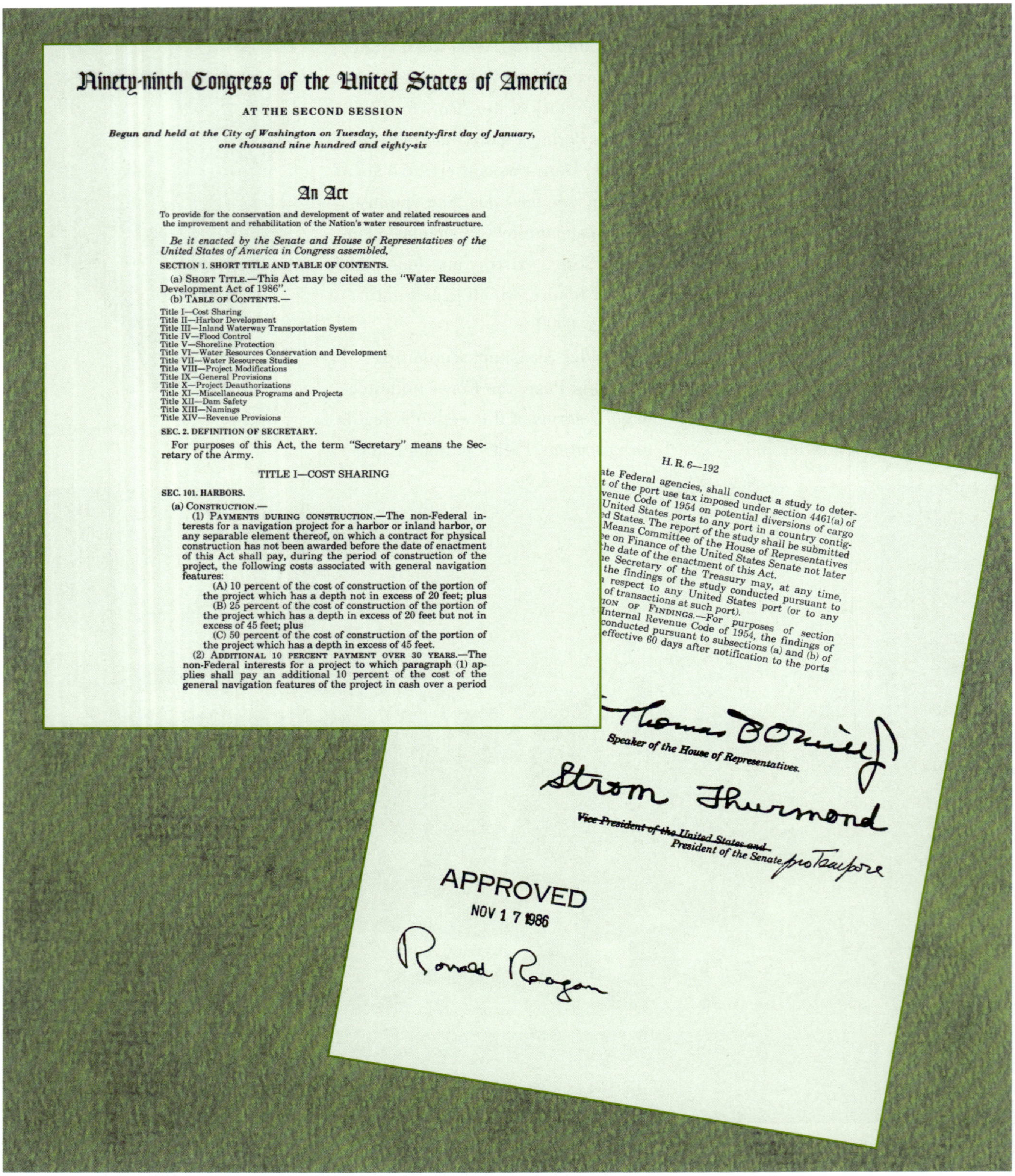

# Ninety-ninth Congress of the United States of America

AT THE SECOND SESSION

*Begun and held at the City of Washington on Tuesday, the twenty-first day of January, one thousand nine hundred and eighty-six*

## An Act

To provide for the conservation and development of water and related resources and the improvement and rehabilitation of the Nation's water resources infrastructure.

*Be it enacted by the Senate and House of Representatives of the United States of America in Congress assembled,*

**SECTION 1. SHORT TITLE AND TABLE OF CONTENTS.**

(a) SHORT TITLE.—This Act may be cited as the "Water Resources Development Act of 1986".

(b) TABLE OF CONTENTS.—



**SEC. 2. DEFINITION OF SECRETARY.**

For purposes of this Act, the term "Secretary" means the Secretary of the Army.

TITLE I—COST SHARING

**SEC. 101. HARBORS.**

(a) CONSTRUCTION.—

(1) PAYMENTS DURING CONSTRUCTION.—The non-Federal interests for a navigation project for a harbor or inland harbor, or any separable element thereof, on which a contract for physical construction has not been awarded before the date of enactment of this Act shall pay, during the period of construction of the project, the following costs associated with general navigation features:

(A) 10 percent of the cost of construction of the portion of the project which has a depth not in excess of 20 feet; plus

(B) 25 percent of the cost of construction of the portion of the project which has a depth in excess of 20 feet but not in excess of 45 feet; plus

(C) 50 percent of the cost of construction of the portion of the project which has a depth in excess of 45 feet.

(2) ADDITIONAL 10 PERCENT PAYMENT OVER 30 YEARS.—The non-Federal interests for a project to which paragraph (1) applies shall pay an additional 10 percent of the cost of the general navigation features of the project in cash over a period

H. R. 6—192

ate Federal agencies, shall conduct a study to deter-
t of the port use tax imposed under section 4461(a) of
venue Code of 1954 on potential diversions of cargo
United States ports to any port in a country contig-
ed States. The report of the study shall be submitted
Means Committee of the House of Representatives
ee on Finance of the United States Senate not later
he date of the enactment of this Act.
e Secretary of the Treasury may, at any time,
the findings of the study conducted pursuant to
respect to any United States port (or to any
of transactions at such port).
ION OF FINDINGS.—For purposes of section
Internal Revenue Code of 1954, the findings of
conducted pursuant to subsections (a) and (b) of
effective 60 days after notification to the ports

Thomas P. O'Neill Jr.
*Speaker of the House of Representatives.*

Strom Thurmond
~~Vice President of the United States and~~
*President of the Senate* pro Tempore

APPROVED
NOV 17 1986
Ronald Reagan

*H. Martin Lancaster*
*Department of Defense*

*Joseph W. Westphal*
*Department of Defense*

changes in interagency agreements gave the Corps more authority in regulating the dredge-and-fill program assigned to the agency in the 1972 Clean Water Act. The Corps also adopted strict time frames and guidelines governing other agencies' input to permit actions and also ensured that the agencies used the same definitions and standards to determine wetland jurisdictions.

With the change in administrations in January 1993, Dorn left office. After a prolonged period in which Acting Assistant Secretaries served, H. Martin Lancaster became the first Assistant Secretary of the Army, Civil Works in the Clinton administration. Lancaster sought to reduce the time and cost of Corps studies and expand engineering and construction management opportunities for the Corps through its reimbursable Support for Others Program. Lancaster, himself a former member of Congress from North Carolina, improved communications with Congress and provided consistent support for the administration's environmental initiatives, especially the restoration of the Everglades and south Florida ecosystem.

Joseph W. Westphal served as the next confirmed Assistant Secretary of the Army, Civil Works from June 1998 to March 2001. He brought a wealth of academic, legislative staff, and executive branch experience to the position. Westphal was a major driving force behind more comprehensive basinwide planning efforts, a revitalization of the

◀ *Water Resources Development Act of 1986*

*Mike Parker*
*Department of Defense*

*John Paul Woodley, Jr.*
*Department of Defense*

Corps' recreation facilities, and an expansion of the Corps' ability to serve the Nation in public infrastructure and environmental restoration needs. His eventual successor, Mike Parker, a former representative from Louisiana, was a strong advocate for maintaining funding levels for Corps programs, but he remained in office for only six months before resigning.

Under Secretary of the Army Les Brownlee subsequently also served as the Acting Assistant Secretary of the Army, Civil Works and then as Acting Secretary of the Army. In 2003 President George W. Bush nominated John Paul Woodley, Jr., as the next Assistant Secretary. Woodley previously held the office of the Assistant Deputy Under Secretary of Defense (Environment), and was principal advisor to the Secretary of Defense on environmental, safety, and occupational health policy and programs. Woodley served in a recess appointment as Assistant Secretary of the Army, Civil Works from August 2003 to December 2004. In May 2005 the Senate confirmed his nomination as assistant secretary. Woodley focused

*Acting Secretary of the Army Les Brownlee meeting with Brig. Gen. Robert Crear, Iraq, September 2003*

the Corps on enhancing performance measurements, streamlining the regulatory process, building planning capabilities, and improving strategic communications.

## Civil Works and the Nation

U.S. Army policy on civil works has continued to stress the need for maximizing the benefits of Corps project investments for the Nation. A notable achievement in this regard was the release by the Corps of its final environmental impact statement on the operation of the Missouri River dams and reservoirs, the *Master Water Control Manual,* after nearly a decade and a half of study. Furthermore, the Corps' plan for restoration along Louisiana's coastal areas also was designed to prioritize the most promising and beneficial remedial activities. The Comprehensive Everglades Restoration Plan to capture, store, and redistribute fresh water previously lost to tides and to regulate the quality, quantity, timing, and distribution of water flows throughout south and central Florida, devised by the Corps and its partners and approved in the Water Resources Development Act of 2000, resulted in a massive ongoing effort to restore the Florida ecosystem. Most recently, the water resources, environmental, regulatory, and emergency response expertise developed through the civil works program has been called upon to support reconstruction efforts in Afghanistan and Iraq.

Acting through the Assistant Secretary's office, the Secretary of the Army has assumed leadership of the Corps' civil works program. The principal responsibility of this position remains overall supervision of the functions of the Department of the Army relating to all aspects of the civil works program, and in specific terms to see that the ongoing and future efforts of the Corps are environmentally sustainable, economically responsible, and fiscally sound. Although form and style have varied according to the political orientation of any given administration, the policies of the Assistant Secretaries of the Army, Civil Works have ensured that the Corps remains the flexible, competent engineering organization that has continuously served the country for two centuries in peace and war.

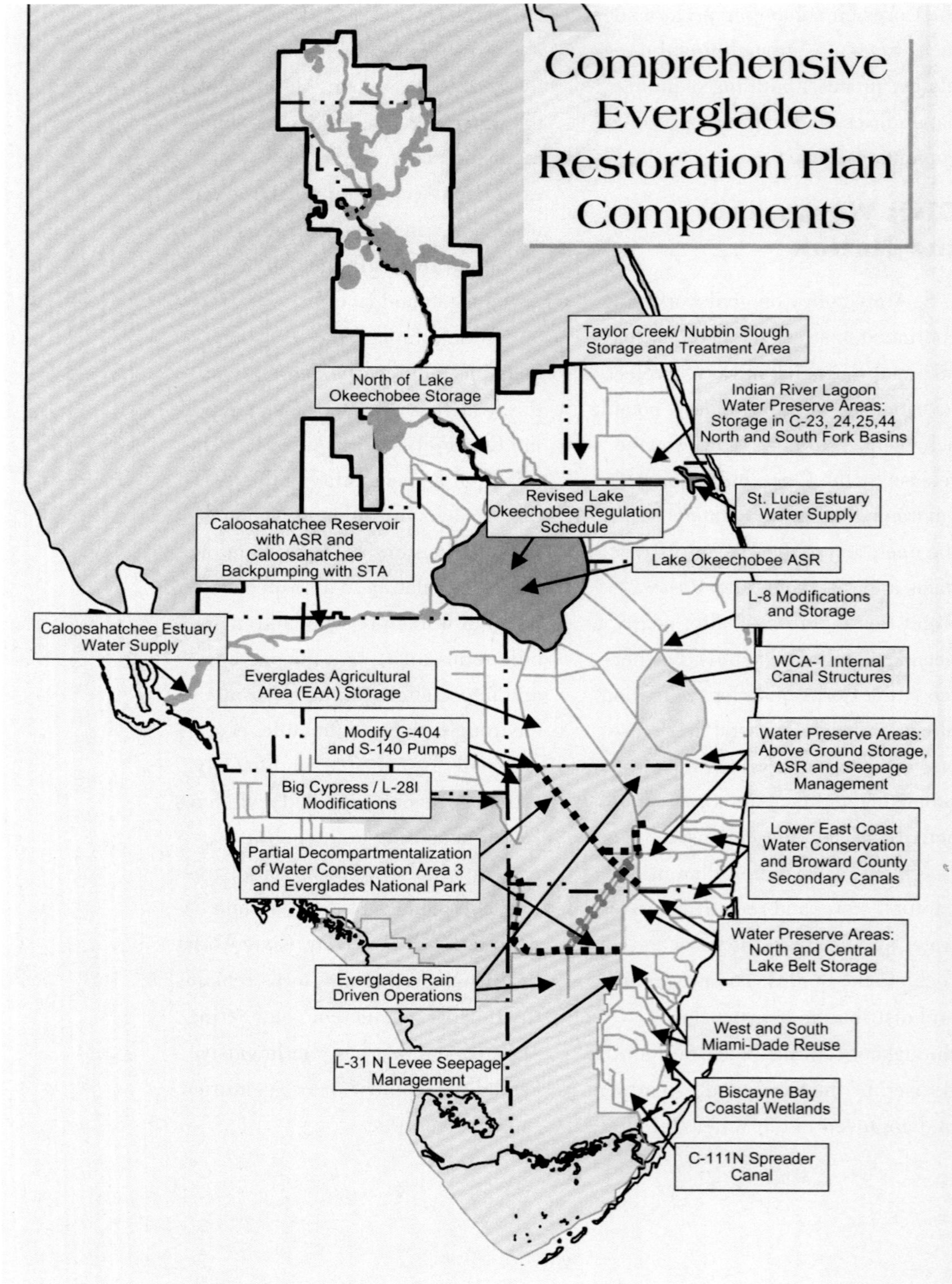


*Comprehensive Everglades Restoration Plan*

# This Chief of Engineers Established the U.S. Army Corps of Engineers as a Partner in the Conservation Movement

More than a decade before the environmental movement took hold, Lieutenant General Samuel D. Sturgis, Jr., Chief of Engineers from 1953 to 1956, envisioned the U.S. Army Corps of Engineers as a partner in the fight for conservation. In a hallmark address to the International Association of Game, Fish, and Conservation Commissioners in September 1953, Sturgis set Corps policy firmly down a path from which it has yet to retreat.

"We must obey the laws of nature and work in harmony with natural forces rather than against them," Sturgis declared in the speech. "Man cannot dominate these forces; but, by working in harmony with them, he can preserve the heritage of future generations." Sturgis traced his own love of nature to his boyhood. All forms of conservation interested him, from soil to wildlife. The destruction of forests filled him with "real pain," and he regretted that in the march "of what we often inaccurately term 'civilization,' some values are likely to be lost."

But General Sturgis believed that the U.S. Army Corps of Engineers could help. The Corps could provide shelter for wildlife on coastal and inland waters, for instance. In fact, Corps projects already furnished "more than 3.5 million acres of land for some form of wildlife management, and recreation." And Sturgis had a vision—namely, to see "resting grounds for migratory game, refuges, managed public hunting, fish culture, game management, research laboratories, field headquarters for wildlife research and administration, arboretums," all aimed at "public use and enjoyment of wildlife resources." Sturgis proclaimed the support of his command toward this cause: "The Corps stands ready and willing to join with each of you and give you every possible assistance that our authorized functions permit to obtain the greatest practicable benefits for wildlife from our projects."

*Lt. Gen. Samuel D. Sturgis, Jr., 1953–56, as Chief of Engineers*

*Pelicans on Gaillard Disposal Island, Mobile Bay, Ala.*

*Damage assessment following the Loma Prieta earthquake, Oakland, Calif., October 1989*

# Emergency Operations

The U.S. Army Corps of Engineers received its first formal federal relief assignment in the winter of 1882 when Mississippi River floods forced thousands of people from their homes. When the Army Quartermaster Department was unable to deliver relief supplies to the shivering refugees, Congress turned to the Corps of Engineers and soon engineer vessels were steaming up and down the river dispensing hundreds of tons of supplies and plucking survivors off rooftops and levees.

In the first half of the twentieth century the Corps' role in providing disaster relief stemmed largely from its flood control responsibilities. The Flood Control Act of 1917 established that flood control was a responsibility of the federal government and placed it under the jurisdiction of the Corps of Engineers. A decade later, during the Mississippi River floods of 1927, the Corps of Engineers organized a massive effort to reinforce the levees to hold back the raging water, but eventually the levees failed, killing hundreds of people and leaving hundreds of thousands homeless. With much of the countryside under water the Corps quickly transitioned its efforts from fighting the flood to helping the communities affected by the disaster. The engineers' relief operations included ferrying supplies to the communities cut off by the rising water and rescuing thousands of beleaguered refugees.

The Corps of Engineers' role in providing disaster relief broadened considerably when Congress passed the landmark Federal Disaster Relief Act of 1950. The act provided a

*Residents of Hickman, Ky., find refuge on levees and rooftops, 1912*

mechanism for local and state governments to request federal assistance, and after determining that a major disaster had indeed occurred, the president could authorize federal agencies to provide "equipment, supplies, facilities, personnel, and other resources" for the preservation of life and property. Additional congressional action followed a series of hurricanes that buffeted the East Coast beginning in 1954. Under PL 84-99 (1955), Congress authorized the Chief of Engineers to undertake activities including disaster preparedness, emergency operations, rehabilitation of flood control works threatened or destroyed by flood, and protection or repair of federally authorized shore protective works threatened or damaged by coastal storms.

*A Corps of Engineers disaster survey team inspects damaged homes following Tropical Storm Agnes.*

Under the provisions of the expanded legislation the Corps was well positioned to lend a helping hand when a string of devastating hurricanes struck the Gulf Coast in the 1960s. In 1965 Hurricane Betsy inundated much of the city of New Orleans, and in 1969 Hurricane Camille came ashore in Mississippi accompanied by a twenty-four-foot storm surge that killed hundreds. In the wake of Hurricane Betsy the Corps helped pump flood waters out of the city, repaired levees, and removed debris. After Hurricane Camille the Corps of Engineers helped clear roads and conducted extensive dredging operations to clear harbors blocked by the storm. In 1972 Tropical Storm Agnes buried much of the east coast under torrential rains that killed more than 100 people and caused more than $3 billion in damage. To cope with the devastation along the eastern seaboard brought on by the storm, the Corps established the Susquehanna District to help house the displaced residents, clear debris, and help make the battered communities livable once again.

The federal government's disaster policy changed again in the 1980s when Congress passed the Robert T. Stafford Disaster Relief and Emergency Assistance Act. The new law tasked the Corps to provide disaster relief support to the newly created Federal Emergency Management Agency (FEMA). That support arrangement was tested in 1992 when Hurricane Andrew roared ashore in South Florida, cutting a twenty-two mile path of devastation from Biscayne Bay to the Everglades. Relief operations in south Florida demonstrated a new level of federal commitment to disaster response: In the months following the disaster, the Corps of Engineers spent nearly $400 million in federal funds installing temporary roofs on some 22,500 homes, removing millions of cubic yards of debris, installing emergency generators and pumps, distributing water, installing temporary housing, and helping rehabilitate nearly 270 schools.

The litany of hurricanes continued —following Hurricane Isabel in 2003, nearly 300 Corps of Engineers personnel deployed to the mid-Atlantic region to distribute water and ice, install generators, and erect more than 100 trailers for temporary housing. In 2004 several hurricanes struck the Gulf Coast and in their wake the Corps' "blue roof program," so named for the color of its distinctive blue plastic coverings, installed 135,000 temporary roofs on homes and businesses across the Gulf region.

In 2005 two powerful hurricanes, Katrina and Rita, struck the Gulf Coast within weeks of one another. High winds and a powerful storm surge inundated much of the city of New Orleans and caused widespread damage across large portions of Louisiana and Mississippi. The Corps' response to the powerful storms was unprecedented; during the relief and recovery efforts more than 3,000 personnel were deployed to the battered communities along the Gulf Coast to assist with relief and recovery operations. Working

*Corps of Engineers contractors install a temporary roof on a home damaged by Hurricane Andrew.*

*Military personnel repair the roof of a school damaged by Hurricane Andrew's strong winds.*

*At a collection point outside of New Orleans, contractors process debris from Hurricane Katrina, October 2005.*

*Corps of Engineers personnel supervising the placement of a community health facility in Chalmette, La., October 2005.*

under the auspices of FEMA and the National Response Plan, the Corps of Engineers mobilized thousands of contractors who removed approximately fifty million cubic yards of debris, installed 193,000 temporary roofs and 914 generators, and repaired more than 1,000 critical public buildings including schools and hospitals.

Operations in and around the city of New Orleans posed special challenges. First, engineers assisted in removing the flood waters from the city. The Corps then launched a crash program to rebuild the city's shattered hurricane protection system to be operational by the start of the 2006 hurricane season.

In addition to hurricanes, during the past century the Corps of Engineers has responded to a variety of other natural disasters including earthquakes and tornados. Following the San Francisco earthquake in 1906, soldiers of the First Battalion of Engineers were the first federal troops to enter the city, and in the weeks that followed they helped feed and house the city's stricken populace and bring raging wildfires under control. When a powerful earthquake rocked south-central Alaska in 1964, the Corps helped remove debris and restore critical municipal services. Following the Loma Prieta, California, earthquake in 1989, and the Northridge, California, earthquake five years later, the Corps provided similar services.

A very different calamity occurred in 1953 when a powerful tornado struck Waco, Texas, killing 114 people and devastating much of the city. Soon after the storm, response personnel from the Fort

San Francisco, Calif., following the April 1906 earthquake and fire

Worth District arrived, set up portable generators, established communications, and within thirty-six hours completed structural assessments of more than 2,000 homes and businesses.

The Corps also has responded to man-made disasters. In 1947 the Galveston District helped evacuate the dead and injured when a devastating explosion destroyed much of Texas City, Texas, killing 500 people and injuring thousands more. In 1989 the tanker *Exxon Valdez* ran aground in Alaska's Prince William Sound, releasing a massive oil spill that threatened large portions of the Alaskan coastline. As government and industry searched for a way to clean up the spill, the Corps modified two of its dredges to vacuum the oil from the water's surface.

Despite more than a century of experience in dealing with disasters and their aftermath, the Nation recoiled in horror when terrorists attacked the World Trade Center and the Pentagon on the morning of September 11, 2001. Soon after the attack in New York, harbor maintenance and survey vessels from the New York District began evacuating 3,000 stranded New Yorkers from lower Manhattan. After discharging

Corps of Engineers personnel confer with a member of the New York City Fire Department at the World Trade Center, September 2001.

their passengers in New Jersey, Corps workboats carried emergency personnel, relief supplies, and fuel back to the city to sustain rescue operations at the World Trade Center. In support of the City of New York and FEMA, the Corps of Engineers brought in mobile command and communication centers to aid emergency operations at the site of the collapsed Trade Center towers. At the same time Corps search and rescue teams searched for survivors while structural engineers assessed the extent of the damage and monitored the condition of the buildings around the World Trade Center complex. The 249th Engineer Battalion also deployed to New York City to help restore power to lower Manhattan and conduct site assessments in and around Wall Street.

The Corps of Engineers was also instrumental in removing and inspecting the nearly 1.6 million tons of debris that resulted from the collapse of the World Trade Center. The Corps and its contractors moved the debris from Manhattan by barge and transported it to the Fresh Kills Landfill on nearby Staten Island. At the landfill the debris was carefully inspected to identify human remains and recover evidence related to the attack and the collapse of the towers. Scores of victims who perished at the World Trade Center were identified on the basis of material recovered during the inspection process.

The terrorist attacks of September 11th placed new emphasis on domestic security, and in December 2002, the Headquarters, U.S. Army Corps of Engineers, established the Homeland Security

The Corps sent its Deployable Tactical Operations System (DTOS) to the World Trade Center to provide communications for rescue workers.

Office within the Civil Works directorate. The new office oversaw the Corps' emergency management program, has played a leading role in assessing the Nation's critical infrastructure, completed numerous facility protection projects, and developed a new risk assessment methodology for dams.

The Corps of Engineers emergency operations function has evolved significantly since 1882 when engineer workboats first carried supplies to flood victims along the Mississippi. Over the course of the last century the federal government has played a progressively larger role in assisting states and municipalities responding to natural and man-made disasters, and the Corps of Engineers' role in providing relief and recovery support has expanded apace. But even as the Corps' mission has expanded into new areas, the foundation of the Corps' value to the Nation—maintaining a nationwide network of engineer districts and divisions with the ability to rapidly mobilize highly skilled and experienced personnel with long-standing relationships with the Nation's construction industry—has remained unchanged.

*(above and below) Government and contractor personnel used mechanized shakers at the Fresh Kills Landfill, N.Y., to screen the debris from the World Trade Center.*

# The U.S. Army Corps of Engineers' Response to Hurricane Andrew

In the early morning hours of August 24, 1992, Hurricane Andrew roared ashore twenty-five miles south of Miami, Florida, hitting Homestead and other south Dade County communities. The hurricane, which possessed one of the highest wind speeds (reported to be 165 mph, with gusts to 185-190 mph), largest storm surges, and lowest barometric pressures ever recorded in the United States during a hurricane, cut a path of destruction twenty-two miles wide and devastated the area from Biscayne Bay to the Everglades. It leveled thousands of homes and other buildings, destroyed public utilities, ripped up trees, and left millions of cubic yards of debris. Its fierce winds tore down most of south Florida's power lines, leaving 1.4 million customers without electricity. After crossing the Florida peninsula and the Gulf of Mexico, it hit southern Louisiana the next day.

The South Atlantic Division and the Jacksonville District of the U.S. Army Corps of Engineers responded immediately, under the overall guidance of the Federal Emergency Management Agency (FEMA). During the next several months the Corps would use almost $400 million in federal funds to help south Florida recover from the devastation.

The Corps provided for immediate human needs. It supplied 5,400 portable toilets to the area and provided hundreds of shower facilities and washers and dryers. Left without a safe water supply, south Floridians relied on the Corps for thousands of gallons of water a day until local water supplies were repaired. With thousands of people homeless, FEMA tasked the Corps to acquire property, clear debris, provide utilities, and put trailers in two large mobile home parks. Corps contractors spent $20 million establishing the parks with more than 250 travel trailers to provide temporary housing.

The Corps also helped to restore vital services to the affected areas. It

(right) Unloading roofing material, Cutler Ridge, Fla.

(below) Temporary housing, Gould, Fla.

turned to its Prime Power units, later organized into the 249th Engineer Battalion, to provide emergency power. In addition to installing twelve of its own 750-kW generators, the Army engineer units supervised the installation of generators and pumps by commercial firms. Prime Power specialists also spearheaded the repair of the Dade County telephone, water, and wastewater treatment systems. Damaged homes needed temporary roof repairs. The Corps and its contractors ultimately supplied 55 million square feet of roofing material and installed it on 22,000 homes. Furthermore, what amounted to a collection of thirty years' worth of debris and refuse littered south Florida in the aftermath of Andrew. Massive amounts of debris blocked roads and posed health problems. The Corps began debris removal quickly. At the peak of debris removal efforts, Corps contractors and troops from the 20th Engineer Brigade operated 2,000 trucks a day. One important mission that involved a remarkable degree of cooperation among agencies was the refurbishment of schools in the devastated areas. A team of Corps personnel, contractors, Navy Seabees, Canadian military personnel, and others opened 268 of Dade County's 278 schools on September 14, only three weeks after Andrew had ripped through the area.

*Corps and contractor personnel install temporary roofing, Perrine, Fla.*

In human terms, Hurricane Andrew was one of the Nation's most debilitating natural disasters, killing twenty people and leaving a quarter of a million homeless. In economic terms, it was one of America's most costly hurricanes, resulting in $26.5 billion in damages. Although the U.S. Army Corps of Engineers was only one actor in the complex drama of south Florida's recovery, the Corps' wealth of experience and its prompt response gave it a leading role in helping the people of the region recover from Andrew's wrath.

# Appendices

# Insignia of the U.S. Army Corps of Engineers

## Corps Castle

The traditional Corps turreted castle is a highly stylized and conventionalized form without decoration or embellishment. There is no evidence that it was patterned after an actual structure. The castle was associated with one of the Corps' earliest responsibilities, the construction of coastal defense fortifications. Some of these early fortifications were called castles. U.S. Military Academy cadets wore the castle emblem as early as 1839 when West Point was part of the Corps of Engineers. In 1840 the Chief Engineer recommended that the castle appear on engineer officers' epaulettes and belt plates. Army regulations first prescribed the use of the castle on engineer caps in 1841. Subsequently the castle has appeared on collar ornaments, shoulder knots, saddle cloths, buttons, and now appears as branch insignia on the dress uniforms of engineer officers and enlisted personnel. Although its design has changed over time, the castle has remained since its inception the distinctive symbol of the Corps of Engineers.

## Essayons Button

As the U.S. Army Corps of Engineers' oldest and most time-honored insignia, the Essayons button has not changed since its first definitely known use during the War of 1812. It is still the required button for the engineer officers' dress uniforms. It is difficult to determine the early history of the castle and the button because the building containing the earliest West Point and Corps of Engineers records burned in 1838. However, early Army records mention "the button of the Engineers" and its already existing device and motto. When the Army prescribed new uniforms by General Orders 7 on February 18, 1840, it described the button as "an eagle holding in his beak a scroll with the word, 'Essayons,' a bastion with embrasures in the distance surrounded by water and a rising sun." Like the castle, the bastion with embrasures symbolized the coastal fortification responsibilities of the Corps. In 1902 when the Army adopted a standard regulation button, it allowed only the Corps of Engineers to retain its own distinctive Essayons button in recognition of the traditions it represented.

## Coat of Arms

In 1867, the U.S. Army Corps of Engineers adopted this Coat of Arms that incorporated the emblems of the Corps of Engineers and the Corps of Topographical Engineers, which had been reunited during the Civil War. This legacy symbol is used primarily for awards, plaques, and honorific presentations related to the military functions of the Corps.

## U.S. Army Engineer School Distinctive Unit Insignia

The United States Army Engineer School, part of the Army Training and Doctrine Command, develops, trains, and supports the engineer force to provide maneuver engineering, force support engineering, and geospatial engineering to Army, Joint, Interagency, and Combined Operations. In 1988, the Engineer School moved to Fort Leonard Wood, Missouri. Personnel assigned to the Army Engineer School are authorized to wear this emblem as a dress uniform device.

## Regimental Distinctive Insignia

The entire U.S. Army Corps of Engineers, as a branch of the Army, is a regiment in the Army's regimental system. The system is designed to enhance loyalty and commitment, esprit de corps, and combat effectiveness. Established in 1986, the regiment officially includes engineer officers and enlisted personnel and civilian employees throughout the Army. The regiment also is closely connected to retired engineer soldiers and civilians and their families. Engineer officers and enlisted personnel wear the regimental insignia on their dress uniforms.

## U.S. Army Corps of Engineers Shoulder Sleeve Insignia

Although associated with the Corps of Engineers becoming a major Army command in 1979, the shoulder sleeve insignia was actually approved for wear by military personnel serving in the Corps' divisions, districts, and other field organizations in 1977 as a way of recognizing those who performed the Corps' military construction, civil works, and other distinctive missions. From 1979 to 2006 the shoulder sleeve insignia was the distinctive component of the Corps' major Army command flag.

## U.S. Army Corps of Engineers Distinctive Unit Insignia

Designed to distinguish the U.S. Army Corps of Engineers when it became a major Army command on June 16, 1979, this insignia incorporated the traditional Corps motto, "Essayons," and a stylized castle above a globe symbolizing the Corps' world-wide responsibilities. It was expected that this distinctive unit insignia would remain unchanged when USACE transitioned from a major Army command to a direct reporting unit in 2006.

## De Fleury Medal

The de Fleury Medal is an award of the Engineer Regiment given to individuals who have made significant contributions to Army engineering. Awarded at the bronze, silver, and gold levels, the medal honors the heroic actions of Revolutionary War engineer François Louis Tesseidre de Fleury at the Battle of Stony Point in July 1779. A French engineer in the service of the Washington's Continental Army, de Fleury led the American troops after his superiors were wounded in recapturing the important position on the Hudson River from the British. A few months later the Continental Congress ordered a medal to be struck honoring de Fleury and that medal was the inspiration for the Engineer Regiment's de Fleury Medal.

## Traditional Castle

Based on the historic Corps castle emblem, this official graphic is authorized for use in special and limited circumstances that call for a sense of the Corps' traditions and history. Since November 30, 1993, it has been a registered trademark of the U.S. Army Corps of Engineers.

## Communications Mark

Adopted after the Corps of Engineers became a major Army command in 1979, this official red and white graphic based on the traditional Corps castle is the standard identifying symbol of the U.S. Army Corps of Engineers. It became a registered trademark of the Corps on November 30, 1993.

**US Army Corps of Engineers®**

## Sapper Tab

The term "sapper" is historically associated with soldiers from the seventeenth and eighteenth centuries who performed the extremely dangerous work of digging trenches toward enemy fortifications during sieges. Approved in 2004, the Sapper Tab is worn on the left shoulder of soldiers who have completed a special Sapper Leaders Course at the U.S. Army Engineer School at Fort Leonard Wood, Missouri. The course emphasizes the role of combat engineers fighting in the front lines with other combat troops.

# Profiles of the Chiefs of Engineers

Since 1775, more than fifty officers have held the highest position among the U.S. Army's engineers. In addition, three officers headed the Topographical Bureau and the Corps of Topographical Engineers between 1818 and 1863. Their likenesses and biographies are on the following pages. Ranks listed are the highest ranks, excluding brevet rank, attained while in office.

## Colonel Richard Gridley

*Chief Engineer, Continental Army (July 1775–April 1776)*

Born January 3, 1710, in Boston, Massachusetts, Richard Gridley was the outstanding American military engineer during colonial warfare with France and served at important battles such as the siege of Louisburg in 1745 and the fall of Quebec in 1759. For his services, he was awarded a commission in the British Army, a grant of the Magdalen Islands, 3,000 acres of land in New Hampshire, and a life annuity. When the break with the mother country came, he stood with the colonies and was made Chief Engineer in the New England Provincial Army. He laid out the defenses on Breed's Hill and was wounded at the Battle of Bunker Hill. He was appointed Chief Engineer of the Continental Army after Washington took command in July 1775. When Washington moved his Army south, Gridley remained as Chief Engineer of the New England Department. He retired in 1781 at age 70. He died June 21, 1796, in Stoughton, Massachusetts.

## Colonel Rufus Putnam

*Chief Engineer, Continental Army (April 1776–December 1776)*

Rufus Putnam was born April 9, 1738, in Sutton, Massachusetts. A millwright by trade, his three years of Army service during the French and Indian War influenced him to study surveying and the art of war. After the Battle of Lexington, he was commissioned an officer of the line, but General Washington soon discovered his engineering abilities. Putnam planned the fortifications on Dorchester Neck that convinced the British to abandon Boston. Washington then brought Putnam to New York as his Chief Engineer. He returned to infantry service in 1777, taking command of the 5th Massachusetts Regiment. He and his troops helped to fortify West Point, erecting strong defenses atop the steep hill that commanded that garrison. The remains of Fort Putnam, preserved by the Military Academy, still honor his name there. Putnam was named a brigadier general in the Continental Army in 1783. In 1788, he led the first settlers to found the present town of Marietta, Ohio. The fortifications that he built there saved the settlements from annihilation during the disastrous Indian Wars. He became surveyor general of federal public lands and judge of the Supreme Court of Ohio. He died in Marietta on May 1, 1824.

## Major General Louis Lebègue Duportail

*Chief Engineer, Continental Army (July 22, 1777–October 10, 1783)*

One of General Washington's most trusted military advisors, Louis Lebègue Duportail, was born near Orleans, France, in 1743. He graduated from the Royal Engineer School in Mézières, France, as a qualified engineer officer in 1765. Promoted to lieutenant colonel in the Royal Corps of Engineers, Duportail was secretly sent to America in March 1777 to serve in Washington's Army under an agreement between Benjamin Franklin and the government of King Louis XVI of France. He was appointed colonel and commander of all engineers in the Continental Army, July 1777; brigadier general, November 1777; commander, Corps of Engineers, May 1779; and major general (for meritorious service), November 1781. Duportail participated in fortifications planning from Boston to Charleston and helped Washington evolve the primarily defensive military strategy that wore down the British Army. He also directed the construction of siege works at Yorktown, site of the decisive American victory of the Revolutionary War. Returning to France in October 1783, Duportail became an infantry officer and in 1788 a field marshal. He served as France's Minister of War during the revolutionary years 1790 and 1791, promoting military reforms. Forced into hiding by radical Jacobins, he escaped to America and bought a farm near Valley Forge, Pennsylvania. He lived there until 1802, when he died at sea while attempting to return to France.

## Lieutenant Colonel Stephen Rochefontaine

*Commandant, Corps of Artillerists and Engineers*
*(February 26, 1795–May 7, 1798)*

Born near Reims, France, in 1755, Stephen Rochefontaine came to America in 1778 after failing to gain a position in the French Royal Corps of Engineers. He volunteered in General Washington's Army on May 15, 1778, and was appointed captain in the Corps of Engineers on September 18, 1778. For his distinguished services at the siege of Yorktown, Rochefontaine was given the brevet rank of major by Congress on November 16, 1781. He returned to France in 1783 and served as an infantry officer, reaching the rank of colonel in the French Army. He came back to the United States in 1792. President Washington appointed him a civilian engineer to fortify the New England coast in 1794. After the new Corps of Artillerists and Engineers was organized, Washington made Rochefontaine a lieutenant colonel and commandant of the Corps on February 26, 1795. Rochefontaine started a military school at West Point in 1795, but the building and all his equipment were burned the following year. He left the U.S. Army on May 7, 1798, and lived in New York City, where he died January 30, 1814. He is buried in old St. Paul's Cemetery in New York.

## Lieutenant Colonel Henry Burbeck

*Commandant, 1st Regiment of Artillerists and Engineers (May 7, 1798–April 1, 1802)*

Born June 8, 1754, in Boston, Massachusetts, Henry Burbeck served as lieutenant of artillery under Colonel Richard Gridley, the Army's first Chief Engineer and artillery commander, in 1775. He remained in the Artillery Corps under General Henry Knox and, in 1777, assumed command of a company of the 3d Continental Artillery Regiment. His unit remained in the North to defend the Hudson Highlands and marched into New York when the British evacuated that city at the close of the Revolutionary War. Honorably discharged in January 1784, Burbeck was reappointed captain of artillery in 1786 and commanded the post at West Point, New York, in 1787–1789. He commanded the U.S. Army's Battalion of Artillery and served as General Anthony Wayne's chief of artillery in the Northwest in 1792–1794. He commanded at Fort Mackinac in 1796–1799. From 1798 to 1802, Burbeck was the senior regimental commander of artillerists and engineers. He also commanded the Eastern Department of the U.S. Army in 1800 and in that year endorsed the creation of a corps of engineers separate from the artillerists. He was chief of the new Artillery Corps from 1802 to 1815, first as a colonel and then, during the War of 1812, as a brevet brigadier general. During the Jefferson administration, Burbeck successfully developed and tested domestically produced cast-iron artillery pieces. He left the Army in June 1815 and died on October 2, 1848, in New London, Connecticut.

## Colonel Jonathan Williams

*Chief Engineer (and first Superintendent of West Point) (April 1, 1802–June 20, 1803, vacated 1803–1805, resumed command April 19, 1805–July 31, 1812)*

Jonathan Williams was born May 20, 1750, in Boston, Massachusetts, a grandnephew of Benjamin Franklin. Williams spent most of the period from 1770 to 1785 in England and France, where he assisted Franklin with business affairs and served as a commercial agent in Nantes. He joined the American Philosophical Society in 1788 and published articles on scientific subjects. President Adams appointed Williams a major in the Corps of Artillerists and Engineers in February 1801, and President Jefferson made him the Army's inspector of fortifications and assigned him to lead the new Military Academy at West Point in December 1801. The following year, Jefferson appointed him to command the separate Corps of Engineers established by Congress on March 16, 1802. Williams also became the first officer to hold the title of Superintendent of the U.S. Military Academy. From 1807 to 1812, Williams designed and completed construction of Castle Williams in New York Harbor, the first casemated battery in the United States. He founded the U.S. Military Philosophical Society and gave it its motto, "Science in War is the Guarantee of Peace." He resigned from the U.S. Army in 1812 and was heading a group of volunteer engineers building fortifications around Philadelphia when he was elected to Congress from that city in 1814. He died in Philadelphia on May 16, 1815.

## Colonel Joseph Gardner Swift

*Chief Engineer (July 31, 1812–November 12, 1818)*

Born December 31, 1783, in Nantucket, Massachusetts, Joseph Swift was appointed a cadet by President John Adams and in 1802 became one of the first two graduates of the Military Academy. He constructed Atlantic coast fortifications from 1804 to 1812, and was only 28 years old when he was appointed colonel, Chief Engineer, and Superintendent of the Military Academy in 1812. As Chief Engineer of the Northern Army, he distinguished himself at the Battle of Chrysler's Farm on November 11, 1813. After completing defensive works in New York, Swift was voted "Benefactor to the City" in 1814. He helped to rebuild the burned capitol in Washington, D.C. He also reorganized the academic staff and planned new buildings at the Military Academy. He resigned from the U.S. Army on November 12, 1818, and was appointed surveyor of the Port of New York. He held that customs post until 1827. Swift was also one of the founders of the first New York Philharmonic Society in 1823. As chief engineer for various railroads, he laid the first "T" rail. From 1829 to 1845, Swift worked for the Corps of Engineers as a civilian, improving two harbors on Lake Ontario. He died July 23, 1865, in Geneva, New York.

## Colonel Walker Keith Armistead

*Chief Engineer (November 12, 1818–June 1, 1821)*

Born in Virginia in 1785, Walker Armistead was named a cadet in the Corps of Artillerists and Engineers by President Jefferson in 1801. On March 5, 1803, he became the third graduate of the new Military Academy and was commissioned in the Corps of Engineers. He served as superintending engineer of the defenses of New Orleans and Norfolk. During the War of 1812, he was successively Chief Engineer of the Niagara frontier army and the forces defending the Chesapeake Bay. He was promoted to colonel and Chief Engineer on November 12, 1818. When the U.S. Army was reorganized on June 1, 1821, he became commander of the 3d Artillery. He was brevetted brigadier general in 1828. He commanded the United States troops that opposed the Seminole Indians in Florida in 1840–1841. He died in Upperville, Virginia, on October 13, 1845.

## Colonel Alexander Macomb

*Chief Engineer (June 1, 1821–May 24, 1828)*

Born April 3, 1782, in Detroit, Alexander Macomb entered the U.S. Army as a cornet of light dragoons in 1799 but was discharged in 1800. He returned to the U.S. Army in 1801 as a second lieutenant of infantry and served as secretary of the commission negotiating treaties with the Indians of the Mississippi Territory. He joined the Corps of Engineers in October 1802 as a first lieutenant and superintended construction of a depot, armory, and fortifications in the Carolinas and Georgia. He also wrote a treatise on military law. After rising to lieutenant colonel in the Corps of Engineers in 1810, he was appointed colonel, 3d Artillery, in 1812 and brigadier general in 1814. In the latter year, he commanded the Lake Champlain frontier force that repulsed a larger veteran British army at Plattsburg. He was voted thanks and granted a gold medal by Congress and brevetted major general. In the reorganized U.S. Army, he was appointed colonel and Chief Engineer in 1821. In that position, he administered the start of federal river and harbor improvements. He was elevated to commanding general of the U.S. Army with the rank of major general in 1828. He died June 25, 1841, in Washington, D.C., and was buried with the highest military honors in Congressional Cemetery. Macomb made the earliest known drawing (1807) to resemble the engineer button.

## Colonel Charles Gratiot

*Chief Engineer (May 24, 1828–December 6, 1838)*

Charles Gratiot was born August 29, 1786, in St. Louis, Missouri. President Jefferson appointed him cadet in 1804. He graduated from the Military Academy in 1806 and was commissioned in the Corps of Engineers. He became a captain in 1808 and assisted Alexander Macomb in constructing fortifications in Charleston, South Carolina. He was post commander of West Point in 1810–1811. He distinguished himself as General William Henry Harrison's Chief Engineer in the War of 1812. He served as Chief Engineer in the Michigan Territory (1817–1818) and superintending engineer for the construction of Hampton Roads defenses (1819–1828). On May 24, 1828, Gratiot was appointed colonel of engineers, brevet brigadier general, and Chief Engineer. For ten years, he administered an expanding program of river, harbor, road, and fortification construction. He also engaged in a lengthy dispute with War Department officials over benefits. In 1838, President Van Buren dismissed him for failing to repay government funds in his custody. Gratiot became a clerk in the General Land Office and died May 18, 1855, in St. Louis.

## Brigadier General Joseph Gilbert Totten

*Chief Engineer (December 7, 1838–April 22, 1864)*

Born August 23, 1788, in New Haven, Connecticut, Joseph Totten graduated from the Military Academy and was commissioned in the Corps of Engineers on July 1, 1805. He resigned in 1806 to assist his uncle, Major Jared Mansfield, who was then serving as surveyor general of federal public lands. Totten reentered the Corps of Engineers in 1808 and assisted in building Castle Williams and other New York Harbor defenses. During the War of 1812, he was Chief Engineer of the Niagara Frontier and Lake Champlain armies. He was brevetted lieutenant colonel for gallant conduct in the Battle of Plattsburg. As a member of the first permanent Board of Engineers in 1816, he laid down durable principles of coastal defense construction. He was appointed Chief Engineer in 1838 and served in that position for 25 years. He was greatly admired by General Winfield Scott, for whom he directed the siege of Veracruz as his Chief Engineer during the Mexican War. He was a regent of the Smithsonian Institution and cofounder of the National Academy of Sciences. He died April 22, 1864, in Washington, D.C.

## Major Isaac Roberdeau

*Chief, Topographical Bureau (August 1, 1818–January 15, 1829)*

Isaac Roberdeau was born in Philadelphia, Pennsylvania, on September 11, 1763. He studied engineering in London, returning to America in 1787 to write, survey, and pursue astronomy. In 1791–1792, he assisted Pierre L'Enfant in planning the new federal capital, the future Washington, D.C. For the next two decades, he practiced engineering in Pennsylvania. His work included assisting William Weston on a canal connecting the Schuylkill and Susquehanna rivers. During the War of 1812, he served in the U.S. Army as a major of topographical engineers, employed chiefly on fortifications. After the war, he assisted the Canadian boundary survey. Secretary of War Calhoun appointed Roberdeau in 1818 to head the newly created Topographical Bureau of the War Department. At first, his duties were largely custodial; he prepared returns and maintained books, maps, and scientific equipment. As the nation turned its attention to internal improvement, Roberdeau used his position to promote the civil activities of the topographical engineers. He was brevetted lieutenant colonel in 1823. He died in Georgetown, Washington, D.C., on January 15, 1829.

## Colonel John James Abert

*Chief, Topographical Bureau (January 31, 1829–April 11, 1861)*
*Chief, Corps of Topographical Engineers (July 7, 1838–September 9, 1861)*

Born September 17, 1788, in Frederick, Maryland, John Abert received an appointment as a Military Academy cadet in January 1808. In 1811, he took a position in the War Department in Washington and resigned as cadet. He joined the District of Columbia Militia as a private during the War of 1812 and fought at the Battle of Bladensburg. In November 1814, he was appointed a topographical engineer with the brevet rank of major. He worked on fortifications, surveys, and river and harbor improvements before being appointed Chief, Topographical Bureau, in 1829. Abert headed the Corps of Topographical Engineers from its creation by Congress in 1838 until he retired in 1861. Under his leadership, the Corps of Topographical Engineers improved the navigability of rivers and harbors, particularly in the basins of the Mississippi River and the Great Lakes; conducted a survey of the hydraulics of the Lower Mississippi River; constructed lighthouses and marine hospitals; explored large portions of the West; and conducted military, border, and railroad surveys. Col. Abert died in Washington, D.C., on January 27, 1863.

## Colonel Stephen H. Long

*Chief, Topographical Bureau (September 9, 1861–March 3, 1863)*
*Chief, Corps of Topographical Engineers (December 12, 1861–March 3, 1863)*

Born in Hopkinton, New Hampshire, on December 30, 1784, Stephen Long graduated from Dartmouth in 1809 and was commissioned in the Corps of Engineers in 1814. Brevetted major, topographical engineers, in April 1816, he conducted extensive explorations and surveys in the old Northwest and Great Plains. Long's Peak was named in his honor. He fixed the nation's northern boundary at the 49th Parallel at Pembina, North Dakota, in 1823. He conducted surveys in the Appalachians for the Baltimore and Ohio Railroad and, in 1829, published his *Railroad Manual or a Brief Exposition of Principles and Deductions Applicable in Tracing the Route of a Railroad.* He served for years as chief engineer for improvement of the western rivers, with headquarters in Cincinnati, Louisville, and finally St. Louis. He became Chief, Corps of Topographical Engineers, in 1861. Upon consolidation of the two corps on March 3, 1863, Col. Long became senior officer to the Chief Engineer, Corps of Engineers. He retired that year and died in Alton, Illinois, on September 4, 1864.

## Brigadier General Richard Delafield

*Chief Engineer (April 22, 1864–August 8, 1866)*

Born September 1, 1798, in New York City, Richard Delafield was the first graduate of the Military Academy to receive a merit class standing, ranking first in the Class of 1818. Commissioned in the Corps of Engineers, he was a topographical engineer with the American commission to establish the northern boundary under the Treaty of Ghent. He served as assistant engineer in the construction of Hampton Roads defenses (1819–1824) and was in charge of fortifications and surveys in the Mississippi River Delta area (1824–1832). While superintendent of repair work on the Cumberland Road east of the Ohio River, he designed and built the first cast-iron tubular-arch bridge in the United States. Appointed Superintendent of the Military Academy after the fire in 1838, he designed the new buildings and the new cadet uniform that first displayed the castle insignia. He superintended the construction of coastal defenses for New York Harbor (1846–1855), was a military observer at the siege of Sevastopol, and was again Superintendent of the Military Academy (1856–1861). Delafield was in charge of New York Harbor defenses (1861–1864) and Chief Engineer from 1864 until his retirement in 1866. He died November 5, 1873, in Washington, D.C. The Secretary of War ordered that 13 guns be fired in his memory at West Point.

## Brigadier General Andrew Atkinson Humphreys

*Chief of Engineers (August 8, 1866–June 30, 1879)*

Andrew Humphreys, born November 2, 1810, in Philadelphia, Pennsylvania, was the son and grandson of chiefs of naval construction. His grandfather designed the U.S.S. *Constitution ("Old Ironsides").* Young Humphreys graduated from the Military Academy in 1831 and served as an artillery officer in Florida during the Seminole War. He resigned from the U.S. Army in 1836, but he accepted an appointment as first lieutenant in the new Corps of Topographical Engineers in 1838. He led a survey of the Mississippi River Delta and, in 1854–1861, headed the Office of Pacific Railroad Explorations and Surveys. His cowritten *Report Upon the Physics and Hydraulics of the Mississippi River,* translated into several languages, became a classic in hydraulic literature. Gen. Humphreys, a distinguished Army corps commander in the Civil War, became Chief of Engineers in 1866. He established the Engineer School of Application and oversaw a substantial expansion of the Corps' river and harbor work. Humphreys held a Harvard degree, published Civil War histories, and was cofounder of the National Academy of Sciences. He died December 27, 1883, in Washington, D.C.

## Brigadier General Horatio Gouverneur Wright

*Chief of Engineers (June 30, 1879–March 6, 1884)*

Born March 6, 1820, in Clinton, Connecticut, Horatio Wright graduated second in the Military Academy Class of 1841 and was commissioned in the Corps of Engineers. He superintended construction at Fort Jefferson at Dry Tortugas, 70 miles west of Key West, Florida, 1846–1856. While assistant to the Chief Engineer of the Army, 1856–1861, he was a member of boards to study iron carriages for seacoast guns and the adaptability of the 15-inch gun for ordnance. He cowrote *Report on Fabrication of Iron for Defenses*. From Chief Engineer of a division at the first Battle of Bull Run, he advanced to command the famous Sixth Army Corps, which saved Washington, D.C., from capture in 1864 and spearheaded the final assault on Petersburg and the pursuit of Lee to Appomattox in 1865. He commanded the Department of Texas, 1865–1866, and served as a member on the Board of Engineers for Fortifications and on many river and harbor planning boards until he was appointed Chief of Engineers in 1879. While Wright was Chief of Engineers, engineer officers began a reservoir system at the headwaters of the Mississippi River and initiated the first substantial federal effort to control the river's lower reaches. Gen. Wright retired March 6, 1884, and died July 2, 1899, in Washington, D.C.

## Brigadier General John Newton

*Chief of Engineers (March 6, 1884–August 27, 1886)*

Born August 24, 1823, in Norfolk, Virginia, a city his father represented in Congress for 31 years, John Newton ranked second in the Military Academy Class of 1842 and was commissioned in the Corps of Engineers. He taught engineering at the Military Academy (1843–1846) and constructed fortifications along the Atlantic Coast and Great Lakes (1846–1852). He was a member of a special Gulf Coast Defense Board (1856) and Chief Engineer, Utah Expedition (1858). Though a fellow Virginian, he did not follow Robert E. Lee but stood firm for the Union in the Civil War. Newton helped construct Washington defenses and led a brigade at Antietam. As division commander, he stormed Marye's Heights at Fredericksburg and fought at Gettysburg and the siege of Atlanta. He commanded the Florida districts in 1864–1866. Returning to the Corps, he oversaw improvements to the waterways around New York City and to the Hudson River above Albany. He also had charge of New York Harbor defenses until he was appointed Chief of Engineers in 1884. Newton was famous for blowing up New York's Hell Gate Rock with 140 tons of dynamite detonated on October 10, 1885. He retired from the Army in 1886 and served as commissioner of public works in New York City (1886–1888) and as president of the Panama Railroad Company (1888–1895). He died on May 1, 1895, in New York.

## Brigadier General James Chatham Duane

*Chief of Engineers (October 11, 1886–June 30, 1888)*

James Duane was born June 30, 1824, in Schenectady, New York. His grandfather was a member of the Continental Congress and mayor of New York City. Duane graduated from Union College in 1844 and from the Military Academy in 1848, where he ranked third in his class. He taught practical military engineering there (1852–1854) during the superintendency of Robert E. Lee. Serving with the U.S. Army's company of sappers, miners, and pontoniers for nine years before the Civil War, Duane led its celebrated 1,100-mile march to Utah in 1858 and commanded select engineer troops to guard President Lincoln at his inauguration in 1861. Duane built the first military ponton bridge over the Potomac at Harpers Ferry in 1862, served as Chief Engineer of the Army of the Potomac (1863–1865), and in seven hours in 1864, built the longest ponton bridge of the Civil War (2,170 feet) across the James River. He commanded at Willets Point, New York (1866–1868), and for ten years constructed fortifications along the coast of Maine and New Hampshire. He was president of the Board of Engineers in 1884–1886 and Chief of Engineers in 1886–1888, when he retired. He then became commissioner of Croton Aqueduct, New York. He published the paper, "History of the Bridge Equipage in the United States Army." Gen. Duane died December 8, 1897, in New York City.

## Brigadier General Thomas Lincoln Casey

*Chief of Engineers (July 6, 1888–May 10, 1895)*

Thomas Lincoln Casey was born May 10, 1831, in Sackets Harbor, New York, where his father, Lieutenant Silas Casey (later an assault team leader in the Battle of Chapultepec in the Mexican War and a general in the Civil War), was then assigned. Young Casey graduated first in the Military Academy Class of 1852 and taught engineering there (1854–1859). During the Civil War, he oversaw Maine coastal fortifications, completing the massive Fort Knox on the Penobscot River. After that war, he headed the division in the Office of the Chief of Engineers responsible for engineer troops, equipment, and fortifications. The Corps' most distinguished builder of monuments and public buildings, Casey headed the Office of Public Buildings and Grounds, District of Columbia, from 1877 to 1881. He built the State, War and Navy Department Building, which is now the Eisenhower Executive Office Building, and completed the Washington Monument. The placing of a sturdier foundation under the partially completed Washington Monument (already 173 feet high) was Casey's greatest engineering feat, but his crowning accomplishment was construction of the Library of Congress Building—all but completed when he died suddenly on March 25, 1896. Burial was at the Casey farm in Rhode Island. Gen. Casey was a member of the National Academy of Sciences and the Society of the Cincinnati and an officer of the Legion of Honor of France.

## Brigadier General William Price Craighill

*Chief of Engineers (May 10, 1895–February 1, 1897)*

William Craighill was born on July 1, 1833, in Charles Town, Virginia (now West Virginia). A classmate of Sheridan, Hood, and McPherson, Craighill ranked second in the Military Academy Class of 1853 and was commissioned in the Corps of Engineers. After working on several Atlantic Coast forts, he taught engineering at the Military Academy in 1859–1862. Another Virginian who stood for the Union, Craighill was division and department engineer during the Civil War and worked on the defenses of Pittsburgh, Baltimore, San Francisco, and New York. After that war, he superintended construction of defenses at Baltimore Harbor and Hampton Roads. He headed the Engineer Office in Baltimore Harbor and Hampton Roads, from 1870 to 1895, overseeing river and harbor work in Maryland and parts of Virginia and North Carolina. When the Corps began to build locks and dams on the Great Kanawha River in West Virginia in 1875, Craighill assumed charge there as well. He completed the first of the moveable wicket dams built in the United States, after visiting France to study their use. He became the Corps' first Southeast Division Engineer. Craighill established the camp for the Yorktown surrender celebration, the first of the sanitary type later adapted to U.S. Army camps. He was a member of the Board of Engineers in 1886–1889. He was appointed Chief of Engineers by President Cleveland in 1895. He retired two years later and died January 18, 1909, in Charles Town, West Virginia.

## Brigadier General John Moulder Wilson

*Chief of Engineers (February 1, 1897–April 30, 1901)*

John Wilson was born October 8, 1837, in Washington, D.C. He graduated from the Military Academy in 1860 and was commissioned in the Artillery Corps. He transferred to the Corps of Topographical Engineers in July 1862 and was awarded the Medal of Honor for fighting at Malvern Hill, Virginia, on August 6, 1862. He joined the Corps of Engineers in 1863 and received three brevets for gallant service in Alabama. After the Civil War, Wilson worked on Hudson River improvements and drafted plans for the canal around the Cascades of the Columbia River. He improved the Great Lakes harbors of Oswego, Cleveland, and Toledo. Wilson headed the divisions of the Chief's office pertaining to military affairs for four years, was in charge of public buildings and grounds in Washington during both of the Cleveland administrations, and was Superintendent of the Military Academy in 1889–1893. Before his appointment as Chief of Engineers in 1897, he was Northeast Division Engineer. As Chief of Engineers, he directed the Corps' activities during the Spanish-American War. He retired April 30, 1901, but remained a prominent figure in the cultural life of Washington until his death there on February 1, 1919.

## Brigadier General Henry M. Robert

*Chief of Engineers (April 30, 1901–May 2, 1901)*

Born May 2, 1837, in South Carolina, Henry Robert graduated fourth in the Military Academy Class of 1857. After receiving his commission in the Corps of Engineers, he taught at the Military Academy and then explored routes for wagon roads in the West and engaged in fortification work in Puget Sound. During the Civil War, he worked on the defenses of Washington and Philadelphia. Robert served as Engineer of the Army's Division of the Pacific in 1867–1871. He then spent two years improving rivers in Oregon and Washington and six years developing the harbors of Green Bay and other northern Wisconsin and Michigan ports. He subsequently improved the harbors of Oswego, Philadelphia, and Long Island Sound and constructed locks and dams on the Cumberland and Tennessee rivers. As Southwest Division Engineer from 1897 to 1901, Robert studied how to deepen the Southwest Pass of the Mississippi River. He was president of the Board of Engineers from 1895 to 1901. On April 30, 1901, he was made brigadier general and was appointed Chief of Engineers. He served until May 2, 1901, when he retired from the U.S. Army. He became famous for his *Pocket Manual of Rules of Order,* a compendium of parliamentary law first published in 1876 and better known today as *Robert's Rules of Order.* He died May 1, 1923, in Hornell, New York.

## Brigadier General John W. Barlow

*Chief of Engineers (May 2, 1901–May 3, 1901)*

John Barlow was born in New York City on June 26, 1838, and graduated from the Military Academy in May 1861. He was first commissioned in the Artillery Corps, but transferred to the Topographical Engineers in July 1862. He served with the Battalion of Engineers at Gettysburg and as engineer of a U.S. Army corps in the siege of Atlanta. He supervised the defenses of Nashville and was brevetted lieutenant colonel for his gallant service there in December 1864. From 1870 until 1874, he was General Sheridan's Chief Engineer in the Military Division of the Missouri. During this period, he made scientific explorations of the headwaters of the Missouri and Yellowstone. His detailed reports became guides for settlers. Barlow improved the harbors and defenses of Long Island Sound from 1875 to 1883, executed harbor improvements in northern Wisconsin and Michigan, and worked on the construction of a canal around Muscle Shoals on the Tennessee River. He was the senior American member of the international commission that remarked the disputed boundary with Mexico in 1892–1896. He was subsequently Northwest Division Engineer for four years. On May 2, 1901, he was commissioned brigadier general and appointed Chief of Engineers. The next day, May 3, 1901, he retired from the U.S. Army after 40 years of service. He died February 27, 1914, in Jerusalem, Palestine.

## Brigadier General George Lewis Gillespie, Jr.

*Chief of Engineers (May 3, 1901–January 23, 1904)*

George Gillespie, Jr., was born October 7, 1841, in Kingston, Tennessee. He graduated second in the Class of 1862 at the Military Academy and was commissioned in the Corps of Engineers. As a Southerner who remained loyal to the Union, Gillespie joined the Army of the Potomac in September 1862. He commanded two companies of the engineer battalion that built fortifications and ponton bridges throughout the Virginia campaigns until the Appomattox surrender. He received the Medal of Honor for carrying dispatches through enemy lines under withering fire to General Sheridan at Cold Harbor, Virginia. He was later Sheridan's Chief Engineer in the Army of the Shenandoah and the Military Division of the Gulf. After the Civil War, Gillespie successively supervised the improvement of harbors at Cleveland, Chicago, Boston, and New York. He initiated construction of the canal at the Cascades of the Columbia River and built the famous lighthouse on Tillamook Rock off the Oregon Coast. Gillespie also served on the Board of Engineers and for six years as president of the Mississippi River Commission. He commanded the U.S. Army's Department of the East in 1898. While Chief of Engineers, he was acting Secretary of War in August 1901. He had charge of ceremonies at President McKinley's funeral and at the laying of the cornerstone of the War College Building in 1903. He served as Army Assistant Chief of Staff in 1904–1905 with the rank of major general. Gen. Gillespie retired June 15, 1905, and died September 27, 1913, in Saratoga Springs, New York.

## Brigadier General Alexander Mackenzie

*Chief of Engineers (January 23, 1904–May 25, 1908)*

Born May 25, 1844, in Potosi, Wisconsin, Alexander Mackenzie graduated from the Military Academy in 1864. Commissioned in the Corps of Engineers, he served with the Union Army in Arkansas in 1864–1865. Mackenzie spent six years commanding a company of engineer troops at Willets Point, New York, that experimented in the use of torpedoes in coastal defense. In 1879, he began a sixteen-year stint as Rock Island District Engineer. He built 100 miles of wing dams on the Upper Mississippi River and produced a 4.5-foot channel between St. Paul and the mouth of the Missouri River. Called to Washington in 1895, he became assistant to the Chief of Engineers in charge of all matters relating to river and harbor improvements. He was a member of the general staff corps and War College Board when he was appointed Chief of Engineers in 1904. He retired May 25, 1908, as a major general, but was recalled to active duty in 1917 at age 73 as Northwest Division Engineer serving again in Rock Island, Illinois. Gen. Mackenzie died March 21, 1921, in Washington, D.C.

## Brigadier General William Louis Marshall

*Chief of Engineers (July 2, 1908–June 11, 1910)*

William Marshall was born June 11, 1846, in Washington, Kentucky, a scion of the family of Chief Justice John Marshall. At age 16, he enlisted in the 10th Kentucky Cavalry, Union Army. He graduated from the Military Academy in 1868 and was commissioned in the Corps of Engineers. Accompanying Lieutenant George Wheeler's Expedition (1872–1876), Marshall covered thousands of miles on foot and horseback and discovered Marshall Pass in central Colorado. He oversaw improvements on the Lower Mississippi River near Vicksburg and on the Fox River Canal System in Wisconsin. As Chicago District Engineer from 1888 to 1899, he planned and began to build the Illinois and Mississippi Canal. Marshall made innovative use of concrete masonry and developed original and cost-saving methods of lock canal construction. While he was stationed at New York (1900–1908), his genius further expressed itself on the Ambrose Channel Project and in fortification construction. He then served for two years as Chief of Engineers. He retired June 11, 1910, but his engineering reputation earned him a special appointment from President Taft as consulting engineer to the Secretary of the Interior on hydroelectric power projects. Gen. Marshall died July 2, 1920, in Washington, D.C.

## Brigadier General William Herbert Bixby

*Chief of Engineers (June 12, 1910–August 11, 1913)*

Born December 27, 1849, in Charlestown, Massachusetts, William Bixby graduated first in the Military Academy Class of 1873 and was commissioned in the Corps of Engineers. After serving with the engineer battalion at Willets Point and as assistant professor of engineering at the Military Academy, Bixby graduated with honors from the French *Ecole des ponts et chaussées.* He received the Legion of Honor for assisting French Army maneuvers. Bixby headed the Wilmington, North Carolina, District from 1884 to 1891. He oversaw improvements on the Cape Fear River, modernized the area's coastal forts, and responded to the earthquake that hit Charleston, South Carolina, in 1886. Bixby served next as District Engineer in Newport, Rhode Island. From 1897 to 1902, he oversaw improvements on the Ohio River and its tributaries from Pittsburgh to Cincinnati. After two years in charge of the Detroit District, he became Chicago District Engineer and Northwest Division Engineer. Bixby was president of the Mississippi River Commission in 1908–1910 and 1917–1918. As Chief of Engineers, he oversaw the raising of the battleship *Maine.* He retired August 11, 1913, but was recalled to service in 1917 as Western Division Engineer. He died September 29, 1928, in Washington, D.C.

## Brigadier General William Trent Rossell

*Chief of Engineers (August 12, 1913–October 11, 1913)*

William Rossell was born in Alabama on October 11, 1849, the son and grandson of U.S. Army officers, and he graduated third in the Military Academy Class of 1873. Commissioned in the Corps of Engineers, he served until 1880 at Willets Point and as assistant professor of engineering at the Military Academy. He then engaged in river, harbor, and fortification work in regions around Portland, Maine; Jacksonville, Florida; and Vicksburg, Mississippi. Rossell served in 1891–1893 as the engineer commissioner on the three-member governing board of the District of Columbia. After briefly commanding the Battalion of Engineers, he led Mobile District for six years. He then supervised lighthouse construction and repair in the New York area and, later, Ohio River improvements. He was a member of the Mississippi River Commission from 1906 to 1913, as well as Central Division Engineer in 1908–1909 and Eastern Division Engineer in 1909–1913. After two months serving as Chief of Engineers, Rossell retired October 11, 1913, but was recalled to active service in 1917. He led the Third New York and Puerto Rico Districts and was Northeast Division Engineer. He again retired in 1918. He died October 11, 1919, in Staten Island, New York.

## Brigadier General Dan Christie Kingman

*Chief of Engineers (October 12, 1913–March 6, 1916)*

Born March 6, 1852, in Dover, New Hampshire, Dan Kingman graduated second in the Military Academy Class of 1875 and was commissioned in the Corps of Engineers. He served as an instructor at the Military Academy and as the engineer officer of the U.S. Army's Department of the Platte. In 1883, he also began the construction of roads and bridges in the new Yellowstone National Park. Kingman directed improvements along the Lower Mississippi River in 1886–1890 and received the thanks of the Louisiana legislature for "splendid service rendered" during the 1890 flood. He oversaw harbor and fortification work on Lake Ontario in 1891–1895 and improvements on the Tennessee River in the last half of that decade. In the latter assignment, he initiated planning for federal cost-sharing with private hydroelectric-power investors for a lock and dam built below Chattanooga. Kingman oversaw substantial harbor improvements at Cleveland in 1901–1905 and headed the Corps' Savannah District and Southeast Division in 1906–1913. The Panama Canal was completed while he was Chief of Engineers. He retired March 6, 1916, and died November 14, 1916, in Atlantic City, New Jersey. Gen. Kingman was buried with high military honors in Arlington National Cemetery. Among the pallbearers were Chief of Staff General Hugh L. Scott and two former Chiefs of Engineers, Generals Mackenzie and Bixby.

## Major General William Murray Black

*Chief of Engineers (March 7, 1916–October 31, 1919)*

Born December 8, 1855, in Lancaster, Pennsylvania, William Black graduated first in the Military Academy Class of 1877 and was commissioned in the Corps of Engineers. From 1886 to 1891, Black headed the Jacksonville District, and in 1897–1898, he was the engineer commissioner on the governing board of the District of Columbia. In the Spanish-American War, he was Chief Engineer, 3d and 5th Army Corps. As Chief Engineer under Generals William Ludlow and Leonard Wood (1899–1901), and six years later as advisor to the Cuban Department of Public Works, he modernized Havana's sanitary system. As commandant of the Army Engineer School (1901–1903), Black moved it from Willets Point, New York, to Washington Barracks, D.C. After his return from Cuba in 1909, he was Northeast Division Engineer and chairman of a board to raise the battleship *Maine.* Devoted to training young engineer officers in the art of war, Gen. Black's greatest responsibility came as Chief of Engineers during World War I in mobilizing and training some 300,000 engineer troops for a wide range of military engineering tasks. For this work, he was awarded the Distinguished Service Medal. He retired October 31, 1919, and died September 24, 1933, in Washington, D.C.

## Major General Lansing Hoskins Beach

*Chief of Engineers (February 10, 1920–June 18, 1924)*

Born June 18, 1860, in Dubuque, Iowa, Lansing Beach graduated third in the Military Academy Class of 1882 and was commissioned in the Corps of Engineers. He developed plans for the reconstruction of the Muskingum River locks and dams soon after Ohio ceded the state-built improvements to the federal government in 1887. From 1894 to 1901, he worked on public improvements in the District of Columbia, serving as engineer commissioner there in 1898–1901. As Detroit District Engineer in 1901–1905, he oversaw harbor improvements as far west as Duluth. Beach supervised improvements along the Louisiana Gulf Coast in 1908–1912 and in Baltimore in 1912–1915. He also oversaw the entire Gulf Division in six of those seven years and the Central Division in 1915–1920. In the latter capacity and as Chief of Engineers, he oversaw construction of the huge Wilson Locks and Dam on the Tennessee River. Beach also served on the Mississippi River Commission and the Board of Engineers for Rivers and Harbors. After his four-year tour as Chief of Engineers, he retired on June 18, 1924. After retirement, Gen. Beach served as consulting engineer for various business interests in the United States and Mexico. He was president of the American Society of Military Engineers and a member of the International Water Commission from 1924 to 1930. He died April 2, 1945, in Pasadena, California.

## Major General Harry Taylor

*Chief of Engineers (June 19, 1924–June 26, 1926)*

Born June 26, 1862, in Tilton, New Hampshire, Harry Taylor graduated from the Military Academy in 1884 and was commissioned in the Corps of Engineers. After serving in engineer offices in Wilmington, North Carolina, and New York City, Taylor served from 1891 to 1900 on fortifications and river and harbor construction work in Oregon and Washington. Later he pursued similar work in New England and New York. Transferred to the Philippines, he supervised all fortification work there in 1904–1905. Taylor was District Engineer in New London, Connecticut, in 1906–1911. He then headed the River and Harbor Division in the Office of the Chief of Engineers for five years. During World War I, he served as Chief Engineer of the American Expeditionary Forces in France (mid-1917 to mid-1918) and received the Distinguished Service Medal. He then served for six years as Assistant Chief of Engineers before assuming the top office in the Corps for two years. Wilson Dam was completed while he was Chief. He was a member of the French Legion of Honor. Gen. Taylor retired June 26, 1926. He died January 27, 1930, in Washington, D.C.

## Major General Edgar Jadwin

*Chief of Engineers (June 27, 1926–August 7, 1929)*

Born August 7, 1865, in Honesdale, Pennsylvania, Edgar Jadwin graduated first in the Military Academy Class of 1890 and was commissioned in the Corps of Engineers. He served with engineer troops in 1891–1895 and was lieutenant colonel of the 3d U.S. Volunteer Engineers in the Spanish-American War. After serving as District Engineer at the expanding ports of Los Angeles and Galveston, he was selected by General Goethals as an assistant in the construction of the Panama Canal. Jadwin served in 1911–1916 in the Office of the Chief of Engineers, focusing on bridge and road matters. Upon the United States' entry into World War I in 1917, he recruited the 15th Engineers, a railway construction regiment, and led it to France. He directed American construction and forestry work there for a year and received the Distinguished Service Medal. President Wilson appointed Jadwin to investigate conditions in Poland in 1919. In 1922–1924, Jadwin headed the Corps' Charleston District and Southeast Division. He then served two years as Assistant Chief of Engineers. As Chief of Engineers, he sponsored the plan for Mississippi River flood control that was adopted by Congress in May 1928. Jadwin retired from the Army on August 7, 1929, and was advanced to lieutenant general on the retired list. He died in Gorgas Hospital in the Canal Zone on March 2, 1931.

## Major General Lytle Brown

*Chief of Engineers (October 1, 1929–October 1, 1933)*

Born November 22, 1872, in Nashville, Tennessee, Lytle Brown graduated fourth in the Military Academy Class of 1898 and was commissioned in the Corps of Engineers. He served with engineer troops in Cuba in 1898 at the Battle of San Juan Hill and the siege of Santiago, and in 1900–1902 he was Engineer of the Department of Northern Luzon in the Philippine Islands. Brown oversaw river improvement projects in 1908–1912 as Louisville District Engineer. He commanded the 2d Battalion of Engineers and served as Engineer of Pershing's 1916 punitive expedition into Mexico. Brown headed the War Plans Division of the War Department general staff from May 1918 to June 1919, addressing important U.S. Army policy issues during and immediately after World War I. He received a Distinguished Service Medal. Brown oversaw construction work at the Wilson Dam Hydroelectric Project in 1919–1920. He was assistant commandant of the Army War College and a brigade commander in the Canal Zone before becoming Chief of Engineers. He concluded his military career as commander of the Panama Canal Department (1935–1936). Gen. Brown retired November 30, 1936. He died in Nashville, Tennessee, on May 3, 1951.

## Major General Edward Murphy Markham

*Chief of Engineers (October 1, 1933–October 18, 1937)*

Born July 6, 1877, in Troy, New York, Edward Markham graduated fifth in the Military Academy Class of 1899 and was commissioned in the Corps of Engineers. He served five years with the 2d Battalion of Engineers, including two years in the Philippines and eight months in Cuba, engaging in military mapping and road and bridge construction. He was Memphis District Engineer (1912–1916) and professor of practical military engineering at the Military Academy. During World War I, he served in France as deputy director, Division of Light Railways and Roads (1918), and in Germany as Chief Engineer, Third Army (1919). After returning to the United States, he was Detroit District Engineer (1919–1925) and commandant of the Army Engineer School, Fort Humphreys, Virginia. Markham then served as Great Lakes Division Engineer. After serving as Chief of Engineers, he made a special military survey in the Hawaiian Islands. Gen. Markham retired February 28, 1938. He was New York public works commissioner in 1938 and president of the Great Lakes Dredge & Dock Company in Chicago from 1938 to 1945. He died in Albany, N.Y., on September 14, 1950.

## Major General Julian Larcombe Schley

*Chief of Engineers (October 18, 1937–October 1, 1941)*

Born February 23, 1880, in Savannah, Georgia, Julian Schley graduated from the Military Academy in 1903 and was commissioned in the Corps of Engineers. He and classmate Douglas MacArthur had their first service with the 3d Battalion of Engineers in the Philippines (1903–1904). Schley later served with engineer troops in the United States and Cuba; as an instructor at the Military Academy; as Assistant Engineer, Washington, D.C.; and as New Orleans District Engineer. During World War I, he commanded the divisional 307th Engineers in the St. Mihiel and Meuse-Argonne offensives and was Engineer, 5th Army Corps, during the last two weeks of the latter drive. He received a Distinguished Service Medal. He was Director of Purchase, General Staff, and a member of the War Department Claims Board in 1919–1920. Schley later served four-year tours as Galveston District Engineer; Engineer of Maintenance, Panama Canal; and governor of the Canal Zone. In the last post, he was also military advisor to the Republic of Panama. Schley was commandant of the Army Engineer School in 1936–1937, before assuming the post of Chief of Engineers. He retired September 30, 1941, but was recalled to active wartime duty in 1943 as director of Transportation, Office of the Coordinator of Inter-American Affairs. He died March 29, 1965, in Washington, D.C.

## Lieutenant General Eugene Reybold

*Chief of Engineers (October 1, 1941–September 30, 1945)*

Born February 13, 1884, in Delaware City, Delaware, Eugene Reybold was distinguished as the World War II Chief of Engineers who directed the largest Corps of Engineers in the nation's history. He graduated from Delaware College in 1903. Commissioned in the Coast Artillery Corps in 1908, Reybold was assigned to military housing and coastal defense construction work. Stationed at Fort Monroe throughout World War I, he became commandant of the Coast Artillery School. He transferred to the Corps of Engineers in 1926 and served as District Engineer in Buffalo, New York; Wilmington, North Carolina; and Memphis, Tennessee. In the last assignment, he successfully battled record Mississippi River flood crests. He was Southwestern Division Engineer (1937–1940) and War Department assistant chief of staff, G–4 (1940–1941). Appointed Chief of Engineers shortly before the Pearl Harbor attack, Gen. Reybold directed the Corps' tremendous range of activities throughout the war and was the first officer ever to rank as lieutenant general while Chief of Engineers. He was awarded a Distinguished Service Medal with Oak Leaf Cluster. Reybold retired January 31, 1946, and died November 21, 1961, in Washington, D.C.

## Lieutenant General Raymond A. Wheeler

*Chief of Engineers (October 4, 1945–February 28, 1949)*

Born July 31, 1885, in Peoria, Illinois, Raymond Wheeler graduated fifth in the Military Academy Class of 1911 and was commissioned in the Corps of Engineers. He served with the Veracruz expedition in 1914 and went to France with the divisional 4th Engineers in 1918. He was awarded a Silver Star for actions in the Aisne-Marne campaign, and by the end of World War I, he had assumed command of his regiment with the rank of colonel. Between the two world wars, he served as District Engineer in Newport, Rhode Island; Wilmington, North Carolina; and Rock Island, Illinois. In September 1941, he was appointed chief of the U.S. Military Iranian Mission and in February 1942 was transferred to the China-Burma-India Theater as commanding general of the Services of Supply. In October 1943, he was assigned to Lord Mountbatten's Southeast Asia command as principal administrative officer and deputy supreme commander. Before the end of World War II, he became commander of the India-Burma Theater. He represented the United States at the Japanese surrender in Singapore. As Chief of Engineers, Wheeler initiated construction of the Missouri River Dams projected in the Pick-Sloan Plan. After his military retirement, he worked for the United Nations and the International Bank for Reconstruction and Development on Asian and African development projects. He oversaw the clearing of the Suez Canal in 1956–1957. Wheeler's U.S. Army decorations included the Distinguished Service Medal with two Oak Leaf Clusters and the Legion of Merit. He was also made an honorary Knight of the British Empire. He died February 8, 1974, in Washington, D.C.

## Lieutenant General Lewis A. Pick

*Chief of Engineers (March 1, 1949–January 26, 1953)*

Born in Brookneal, Virginia, November 18, 1890, Lewis Pick graduated from Virginia Polytechnic Institute in 1914. During World War I, he served with the 23d Engineers in France. Pick received his Regular Army commission in the Corps of Engineers on July 1, 1920. He served in the Philippines from 1921 until 1923 and helped organize an engineer regiment composed of Filipino soldiers. He was District Engineer at New Orleans during the great 1927 Mississippi River floods, and he helped coordinate federal relief efforts. Pick was named Missouri River Division Engineer in 1942 and, with W. Glenn Sloan of the Bureau of Reclamation, he cowrote the Pick-Sloan Plan for controlling the water resources of the Missouri River Basin. Pick was assigned to the China–Burma–India Theater of Operations in October 1943, and oversaw the construction of the Ledo Road across northern Burma from India to China. After his return to the United States in 1945, he served again as Missouri River Division Engineer. On March 1, 1949, President Truman appointed him Chief of Engineers. Pick was awarded the Distinguished Service Medal with Oak Leaf Cluster. He died December 2, 1956, in Washington, D.C.

## Lieutenant General Samuel D. Sturgis, Jr.

*Chief of Engineers (March 17, 1953–September 30, 1956)*

Born July 16, 1897, in St. Paul, Minnesota, Samuel Sturgis, Jr., came from an illustrious military family. Both his father and grandfather were Military Academy graduates and major generals. Young Sturgis graduated from the Military Academy in 1918. As a junior engineer officer, he taught mathematics at the Academy for four years. In 1926, he was ordered to the Philippines, where he served as adjutant of the 14th Engineers. His strategic studies of the islands over a three-year period developed knowledge he used later when he returned to the Philippines in 1944 as Chief Engineer of General Walter Krueger's Sixth Army. Sturgis commanded a mounted engineer company at Fort Riley, Kansas, in 1929–1933 and encouraged the adoption of heavy mechanical equipment. He was District Engineer in 1939–1942 in Vicksburg, Mississippi, where he worked on flood control and a large military construction program. In 1943–1945, Sturgis's engineer troops built roads, airfields, ports, and bases from New Guinea to the Philippines. Sturgis was senior engineer for the nation's air forces in 1946–1948 and was Missouri River Division Engineer in 1949–1951. In 1951, he became the commanding general of the 6th Armored Division and Fort Leonard Wood. In 1952, he was appointed commanding general of the Communications Zone, supporting the United States Army in Europe. He became Chief of Engineers on March 17, 1953. His military decorations included the Distinguished Service Medal with Oak Leaf Cluster, Silver Star, Legion of Merit, and Bronze Star. He died July 5, 1964, in Washington, D.C.

## Lieutenant General Emerson C. Itschner

*Chief of Engineers (October 1, 1956–March 27, 1961)*

Born in Chicago, Illinois, July 1, 1903, Emerson Itschner graduated from the Military Academy in 1924 and was commissioned in the Corps of Engineers. He obtained a degree in civil engineering from Cornell University in 1925. Itschner served with the Alaska Road Commission in 1927–1929. He taught at the Missouri School of Mines and served as assistant to the Upper Mississippi Valley Division Engineer and the St. Louis District Engineer. He commanded a topographic survey company in 1940–1941. In 1942–1943, Itschner headed the office in Corps headquarters that supervised U.S. Army airfield construction in the forty-eight states. In 1944–1945, he oversaw the reconstruction of ports and the development of supply routes to U.S. forces in Europe as the engineer for the Advance Section, Communications Zone. Itschner headed the division in Corps headquarters responsible for military construction operations from 1946 to 1949. After a year as Seattle District Engineer, he went to Korea as Engineer of I Corps and oversaw engineer troop operations in western Korea. He was North Pacific Division Engineer in 1952–1953. From 1953 until being appointed Chief of Engineers, he served as Assistant Chief of Engineers for Civil Works. He was awarded the Distinguished Service Medal, Legion of Merit with two Oak Leaf Clusters, Bronze Star, and Purple Heart. Gen. Itschner retired in 1961 and died in Portland, Oregon, on March 15, 1995.

## Lieutenant General Walter K. Wilson, Jr.

*Chief of Engineers (May 19, 1961–June 30, 1965)*

The son of an artillery officer, Walter Wilson, Jr., was born at Fort Barrancas, Florida, on August 26, 1906. He graduated from the Military Academy in 1929 and was commissioned in the Corps of Engineers. Before 1942, he served with troops, continued his military and engineering education, and was an instructor at the Military Academy. During World War II, Wilson served as Deputy Engineer-in-Chief with the Southeast Asia Command at New Delhi, India, and Kandy, Ceylon. He became commanding general, Advance Section, U.S. Forces, India-Burma Theater, and chief of staff of the Chinese Army in India. After the consolidation of Intermediate and Base Sections with Advance Section, Wilson commanded all ground forces remaining in the theater. He was District Engineer in St. Paul, Minnesota (1946–1949), and Mobile, Alabama (1949–1952), and Mediterranean Division Engineer (1953–1955). He assumed command of the 18th Engineer Brigade at Fort Leonard Wood, Missouri, in 1955. He served as Deputy Chief of Engineers for Construction from 1956 to 1960. Wilson was Commanding General, the Army Engineer Center and Fort Belvoir, and Commandant, the Army Engineer School in 1960–1961. He retired as Chief of Engineers on June 30, 1965. Wilson's military honors included the Legion of Merit with Oak Leaf Cluster, the Soldier's Medal, and membership in the French Legion of Honor. He died in Mobile, Alabama, on December 6, 1985.

## Lieutenant General William F. Cassidy

*Chief of Engineers (July 1, 1965–July 31, 1969)*

Born on a U.S. Army post near Nome, Alaska, on August 28, 1908, William Cassidy graduated from the Military Academy in 1931 and was commissioned in the Corps of Engineers. He served as assistant to the District Engineer in Portland, Oregon, commanded an engineer company at Fort Belvoir, Virginia, and oversaw military construction projects in Hawaii. During World War II, Cassidy commanded engineer troops specializing in airfield construction in England, North Africa, and Italy. He was deputy chief, then chief, of the War Plans (later Operations and Training) Division, Office of the Chief of Engineers, in 1944–1947. At the outbreak of the Korean Conflict, he was ordered to Japan, where he was responsible for engineer supply. He served as South Pacific Division Engineer from 1955 to 1958 and was the senior logistics advisor to the Republic of Korea Army in 1958–1959. Cassidy was the Corps' Director of Civil Works from September 1959 to March 1962 and was then appointed Deputy Chief of Engineers. On March 1, 1963, he became the commanding general of the Army Engineer Center and Fort Belvoir, and commandant of the Army Engineer School. Cassidy became Chief of Engineers on July 1, 1965, and held that post for four years. He was awarded the Distinguished Service Medal for his service as Chief of Engineers. His other military decorations included the Legion of Merit with Oak Leaf Cluster, the Bronze Star, and the Republic of Korea Presidential Citation. He died in Longwood, Florida, on March 31, 2002.

## Lieutenant General Frederick J. Clarke

*Chief of Engineers (August 1, 1969–July 31, 1973)*

Born in Little Falls, New York, on March 1, 1915, Frederick Clarke was commissioned in the Corps of Engineers in 1937 after graduating fourth in his Military Academy class. Clarke received a master's degree in civil engineering from Cornell University in 1940 and later attended the Advanced Management Program of the Graduate School of Business, Harvard University. During World War II, he commanded a battalion that helped construct a military airfield on Ascension Island in the South Atlantic, and he served in Washington, D.C., at Headquarters, Army Service Forces. After the war, Clarke worked in the atomic energy field for the Manhattan District and the Atomic Energy Commission at Hanford, Washington, and at the Armed Forces Special Weapons Project at Sandia Base, Albuquerque, New Mexico. As the District Engineer of the Trans-East District of the Corps in 1957–1959, he was responsible for U.S. military construction in Pakistan and Saudi Arabia, and he initiated transportation surveys in East Pakistan and Burma. In the decade before his appointment as Chief of Engineers, Clarke was Engineer Commissioner of the District of Columbia (1960–1963); Commanding General, the Army Engineer Center and Fort Belvoir, and Commandant, the Army Engineer School (1965–1966); and Deputy Chief of Engineers (1966–1969). As Chief of Engineers, Clarke guided the Corps as it devoted increased attention to the environmental impact of its work. Gen. Clarke was awarded the Distinguished Service Medal and the Legion of Merit. He died at Fort Belvoir, Virginia, on February 4, 2002.

## Lieutenant General William C. Gribble, Jr.

*Chief of Engineers (August 1, 1973–June 30, 1976)*

Born in Ironwood, Michigan, on May 24, 1917, William Gribble, Jr., graduated from the Military Academy in 1941 and was commissioned in the Corps of Engineers. During World War II, he served on the staff of the 340th Engineer General Service Regiment as it first built a section of the Alaska Highway in western Canada and later assisted General MacArthur's drive in New Guinea and the Philippines. At the end of the war, he commanded the 118th Engineer Combat Battalion, 43d Infantry Division. Gribble then worked in the Los Alamos laboratory and in the Reactor Development Division of the Atomic Energy Commission. As Alaska District Engineer, he oversaw construction of a nuclear power plant at Fort Greeley, Alaska. He headed the U.S. Army's nuclear power program in 1960–1961. In 1963, he was the Corps' North Central Division Engineer. Gribble's scientific skills led to his service as director of research and development in the Army Materiel Command in 1964–1966 and as the U.S. Army's Chief of Research and Development, 1971–1973. In 1969–1970, he commanded the Army Engineer Center and Fort Belvoir, and was commandant of the Army Engineer School. He became Chief of Engineers in 1973. Gribble received a master's degree in physical science from the University of Chicago in 1948 and an honorary doctorate in engineering from Michigan Technological University. He was also an honorary member of the United Kingdom's Institute of Royal Engineers. His decorations included the Distinguished Service Medal with Oak Leaf Cluster, the Legion of Merit with Oak Leaf Cluster, and the Brazilian Order of Military Merit. Gen. Gribble died at Fort Belvoir, Virginia, on June 2, 1979.

## Lieutenant General John W. Morris

*Chief of Engineers (July 1, 1976–September 30, 1980)*

John Morris was born in Princess Anne, Maryland, on September 10, 1921. He graduated from the Military Academy in June 1943 and was commissioned in the Corps of Engineers. During World War II, he commanded an airfield construction company in the Western Pacific. After the war, he served in the Philippines and Japan, in the Corps' Savannah District, and as area engineer at Goose Bay, Labrador. In 1960–1962, he commanded the divisional 8th Engineer Battalion in Korea. Morris headed the Corps' Tulsa District in 1962–1965 as it improved navigation on the Arkansas River. During the peak years of the Vietnam War, he was the U.S. Army's Deputy Chief of Legislative Liaison (1967–1969), and he commanded the 18th Engineer Brigade in Vietnam (1969–1970). He was then Missouri River Division Engineer for two years, the Corps' Director of Civil Works for three years, and Deputy Chief of Engineers in 1975–1976. As Chief of Engineers, Morris convinced the U.S. Army to include the Corps of Engineers among the major commands. Morris obtained a master's degree in civil engineering from the University of Iowa. His military awards include the Distinguished Service Medal, the Legion of Merit with three Oak Leaf Clusters, the Bronze Star, and the Defense Meritorious Service Medal. Gen. Morris was selected Construction's Man of the Year for 1977 by the *Engineering-News Record.*

## Lieutenant General Joseph K. Bratton

*Chief of Engineers (October 1, 1980–September 14, 1984)*

Joseph Bratton was born on April 4, 1926, in St. Paul, Minnesota. He graduated third in the Class of 1948 at the Military Academy and was commissioned in the Corps of Engineers. He served with an engineer battalion in Austria in 1949–1952 and with the divisional 13th Engineer Combat Battalion in Korea in 1953–1954, both before and after the armistice. He later commanded the 24th Engineer Battalion, 4th Armored Division, in Germany (1964–1965) and the 159th Engineer Group in Vietnam (1969–1970). Bratton also held numerous staff assignments. He was a military assistant to Secretary of the Army Stanley Resor in 1967–1969 and secretary to the Joint Chiefs of Staff in 1970–1972. Having received a master's degree in nuclear engineering from the Massachusetts Institute of Technology in 1959, Bratton served as chief of Nuclear Activities at Supreme Headquarters, Allied Powers, Europe (SHAPE), in 1972–1975 and Director of Military Applications at the U.S. Department of Energy in 1975–1979. His last assignments before becoming Chief of Engineers in October 1980, were as Division Engineer of the Corps' South Atlantic Division (1979–1980) and then briefly as Deputy Chief of Engineers. His military awards include the Defense Distinguished Service Medal, the Army Distinguished Service Medal, the Legion of Merit with two Oak Leaf Clusters, and the Bronze Star with Oak Leaf Cluster.

## Lieutenant General Elvin R. Heiberg III

*Chief of Engineers (September 14, 1984–May 5, 1988)*

Elvin Heiberg III was born at Schofield Barracks, Honolulu, Hawaii, on March 2, 1932. He became a third-generation West Pointer when he graduated fifth in the Military Academy Class of 1953. He later obtained three masters' degrees (civil engineering from the Massachusetts Institute of Technology, and government and administration from George Washington University). Early in his military career, Heiberg served as operations officer of the 3d Brigade, 3d Infantry Division in Germany and taught in the Social Sciences Department at the Military Academy. In 1968–1969, he commanded the divisional 4th Engineer Battalion in Vietnam and was awarded a Silver Star. He then served as special assistant and executive assistant to the director of the Office of Emergency Preparedness. Heiberg served for a year as executive to Secretary of the Army Howard Callaway. He then headed the Corps' New Orleans District and, in 1975–1978, the Ohio River Division. He served as senior engineer on the staff of U.S. Army, Europe, in 1978–1979. Heiberg was the Corps' Director of Civil Works in 1979–1982 and then Deputy Chief of Engineers. After managing the U.S. Army's Ballistic Missile Defense Program for a year, he became Chief of Engineers in 1984. Heiberg graduated from the Industrial College of the Armed Forces. His military awards include the Distinguished Service Medal, the Legion of Merit with two Oak Leaf Clusters, the Distinguished Flying Cross, and the Bronze Star.

## Lieutenant General Henry J. Hatch

*Chief of Engineers (June 17, 1988–June 4, 1992)*

The son of an artillery officer, Henry J. Hatch was born on August 31, 1935, in Pensacola, Florida. After graduating from the U.S. Military Academy in 1957, he completed airborne and ranger training at Fort Benning, Georgia, and earned a master's degree in geodetic science at Ohio State University. Hatch held several leadership positions in U.S. Army airborne and airmobile units early in his career. He commanded a company of the 82d Airborne Division's 307th Engineer Battalion at Fort Bragg, North Carolina; served on the staff of the 2d Airborne Battle Group, 503d Infantry in Okinawa; and commanded the 326th Engineer Battalion of the 101st Airborne Division in Vietnam in 1968–1969. Hatch subsequently oversaw West Point construction work for the Corps' New York District and in 1974 began a three-year tenure as Nashville District Engineer. He then returned to the Far East to lead the 2d Infantry Division Support Command in Korea and later directed U.S. Army and U.S. Air Force construction in Korea, Japan, and the Pacific as Division Engineer of the Corps' Pacific Ocean Division. Hatch was Deputy Chief of Staff, Engineer, for U.S. Army, Europe, in 1981–1984. He next returned to the Corps of Engineers, serving briefly as Assistant Chief of Engineers and then for nearly four years as director of Civil Works. President Reagan nominated him as Chief of Engineers in May 1988. Lt. Gen. Hatch has been awarded the Legion of Merit, two Meritorious Service Medals, two Bronze Stars, three Air Medals, and two Army Commendation Medals.

## Lieutenant General Arthur E. Williams

*Chief of Engineers (August 24, 1992–June 30, 1996)*

Born in Watertown, New York, on March 28, 1938, Arthur Williams obtained a commission as a U.S. Army engineer officer upon his graduation in 1960 from Saint Lawrence University, where he majored in mathematics. He later obtained a bachelor's degree in civil engineering from Rensselaer Polytechnic Institute and a master's degree in civil engineering and economic planning from Stanford University. Williams commanded an armored engineer company in Germany and an engineer construction company in Vietnam. During a second tour in Vietnam, he served as operations officer of the 577th Engineer Battalion. He later commanded the 44th Engineer Battalion in Korea and was an assignment officer at the U.S. Army Military Personnel Center. Williams headed the Corps' Sacramento District in 1982–1985 and then served as Chief of Staff, Corps headquarters. He subsequently headed the Pacific Ocean Division and then the Lower Mississippi Valley Division. He was also president of the Mississippi River Commission. He returned to Corps headquarters in July 1991 as Director of Civil Works. Williams was nominated as Chief of Engineers by President Bush in 1992. His military awards include two Bronze Stars, three Legion of Merit Awards, the Defense Meritorious Service Medal, and the Army Commendation Medal.

## Lieutenant General Joe N. Ballard

*Chief of Engineers (September 18, 1996–August 31, 2000)*

A native of Oakdale, Louisiana, Joe N. Ballard was born on March 27, 1942, and graduated in 1965 from Southern University and A&M College, Baton Rouge, Louisiana, with a degree in electrical engineering. After receiving a commission in the Corps of Engineers, he served as a platoon leader during his first tour in South Vietnam and as a company commander and chief of the lines of communication section of the 18th Engineer Brigade during his second tour. Following assignments with Fifth Army and the Recruitment Command, he was a staff officer in the 326th Engineer Battalion, 101st Airborne Division. Ballard served on the staff of the U.S. Forces, Korea, Engineer, and later was the engineer in the Army Energy Office. In 1982 he went to West Germany as commander of the 82d Engineer Battalion, and later he commanded the 18th Engineer Brigade and served on the staff of the Deputy Chief of Staff, Engineer, U.S. Army, Europe. In 1991 he became assistant commandant of the school and deputy commanding general of the Engineer Center and Fort Leonard Wood. After serving as Chief, Total Army Basing Study, Ballard returned to Missouri as Commanding General of the Engineer Center and Fort Leonard Wood. When he was nominated to be Chief of Engineers he was chief of staff of the Training and Doctrine Command. Ballard earned a master's degree in engineering management from the University of Missouri. His military awards include the Distinguished Service Medal, three Legion of Merit Awards, two Bronze Stars, the Defense Meritorious Service Medal, four Meritorious Service Medals, and two Army Commendation Medals.

## Lieutenant General Robert B. Flowers

*Chief of Engineers (October 23, 2000–July 1, 2004)*

Born in Kane, Pennsylvania, on July 9, 1947, Robert B. Flowers was the son of a military officer. Following graduation from the Virginia Military Institute and commissioning in the Corps of Engineers in 1969, he completed Airborne and Ranger training. In 1976 he earned a master's degree in civil engineering from the University of Virginia. From 1970 to 1972 Flowers was a platoon commander, company commander, and operations officer of the 94th Engineer Battalion in West Germany. He served in the Army Support Command in Thailand and the Portland Engineer District. From 1980 to 1984 Flowers was on the staffs of the 20th Engineer Brigade and 307th Engineer Battalion, 82d Airborne Division, and in 1985 he took command of the 307th. He served on the Joint Staff before taking command of the 20th Engineer Brigade, XVIII Airborne Corps, during Operations Desert Shield and Desert Storm. He was Unified Task Force Engineer, United Nations Task Force, during Operation Restore Hope in Somalia. After a tour as Deputy Assistant and then Assistant Commandant of the Engineer School at Fort Leonard Wood, he went to South Korea in 1994 as Assistant Division Commander of the 2nd Infantry Division. Flowers then became commander of the Lower Mississippi Valley Engineer Division and deployed briefly in 1996 to Bosnia. Prior to his selection as Chief of Engineers, he was Commanding General of the Maneuver Support Center and Commandant of the Engineer School at Fort Leonard Wood, Missouri. He earned two Distinguished Service Medals, four Legion of Merit awards, a Bronze Star, a Defense Meritorious Service Medal, two Meritorious Service Medals, and four Army Commendation Medals.

## Lieutenant General Carl A. Strock

*Chief of Engineers (July 1, 2004–May 17, 2007)*

Born at Fort Benning, Georgia, on July 20, 1948, Carl A. Strock grew up in an Army family. He enlisted in the U.S. Army and received his commission in the Infantry from Officer Candidate School in July 1972. He received a bachelor's degree in civil engineering from the Virginia Military Institute in 1975. After receiving Ranger and Special Forces training, he served primarily in infantry units until 1983 when he transferred to the engineer branch. He was a staff officer in the 307th Engineer Battalion, 82d Airborne Division. After teaching as an exchange officer at the Royal School of Military Engineering, Chattenden, England, he became commander of the 307th Engineer Battalion, leading the unit in Operation Just Cause in Panama and Operations Desert Shield and Desert Storm. He commanded the Engineer Brigade, 24th Infantry Division, at Fort Stewart, Georgia. In 1996 he became Chief of Staff of the Engineer Center and Fort Leonard Wood, Missouri. A year later, he became commander of the Pacific Ocean Engineer Division in Hawaii and then the Northwestern Engineer Division in Portland, Oregon. In the fall of 2001 he became Director of Military Programs in Corps headquarters. In March 2003 he went to Iraq where he was Deputy Director of Operations for the Coalition Provisional Authority for six months.

He returned to headquarters as Director of Civil Works and became Chief of Engineers in July 2004, retiring from the position in June 2007. Strock has a master's degree in civil engineering from Mississippi State University. His awards include a Defense Distinguished Service Medal, two Distinguished Service Medals, two Legion of Merit Awards, two Bronze Stars, three Meritorious Service Medals, and two Army Commendation Medals.

## Lieutenant General Robert L. "Van" Van Antwerp

*Chief of Engineers (May 18, 2007–)*

Robert L. Van Antwerp was born on January 27, 1950, in Benton Harbor, Michigan. He graduated from the U.S. Military Academy in 1972 and completed Ranger, Airborne, and Air Assault training. He served as a platoon leader in the 76th Engineer Battalion (Construction) and executive officer and assistant division engineer of the 65th Engineer Battalion, 25th Infantry Division, in Hawaii before attending the University of Michigan where he received a master of science degree in mechanical engineering in 1981. Later he earned a master of business administration degree from Long Island University. He taught in the Department of Mechanics at West Point for three years. Van Antwerp was executive officer of the 84th Engineer Battalion and chief of the engineering and construction division of the Western Command before serving a year as executive officer, Office of the Chief of Engineers. He became commander of the 326th Engineer Battalion, 101st Airborne Division (Air Assault), in April 1989 and led the battalion in Operations Desert Shield and Desert Storm. He was commander of the Los Angeles Engineer District (1992 to 1994) during the Northridge Earthquake. After a six-month tour as chief of staff, Headquarters, U.S. Army Corps of Engineers, he became executive assistant to the vice chairman of the Joint Chiefs of Staff. In 1996 he went to Atlanta, Georgia, where he was commanding general of the Corps' South Atlantic Division. After a senior assignment in the Army Secretariat, Van Antwerp was named assistant chief of staff for installation management. In July 2002 he became commanding general of the Maneuver Support Center and Fort Leonard Wood. He followed this two-year tour with more than two years as commanding general of the Army Accessions Command, Training and Doctrine Command. In May 2007 Van Antwerp was confirmed as Chief of Engineers. His awards include two Distinguished Service Medals, the Defense Superior Service Medal, two Legion of Merit awards, the Bronze Star, and five Meritorious Service Medals.

# Selected Bibliography


## I. GENERAL

American Public Works Association. *History of Public Works in the United States, 1776–1976*. Edited by Ellis L. Armstrong (Chicago, Ill., 1976).

Baldwin, William C. "History of Peacetime Housing to the End of the Cold War." *Occasional Papers 4. Society for History in the Federal Government* (2004): 1–30.

Crackel, Theodore J. *West Point: A Bicentennial History* (Lawrence, Kan., 2002).

Crowley, Robert, editor. *West Point: The First 200 Years* (New York, N.Y., 2002).

Dunne, David M. "The Engineer School—Past and Present." *The Military Engineer* 41 (November–December 1949): 411–16.

Forman, Sidney. *West Point: A History of the United States Military Academy* (New York, N.Y., 1950).

Harper, Michael W., Thomas R. Reinhardt, and Barry Sude. *Environmental Cleanup at Former and Current Military Sites: A Guide to Research* (Alexandria, Va., 2001).

McCullough, David G. *The Path Between the Seas: The Creation of the Panama Canal, 1870–1914* (New York, N.Y., 1977).

Morrison, James L., Jr. *"The Best School": West Point, 1833–1836* (Kent, Ohio, 1986).

Scott, Pamela. *Capital Engineers: The U.S. Army Corps of Engineers in the Development of Washington, D.C., 1790–2004* (Alexandria, Va., 2005).

Wermiel, Sara. *Army Engineers' Contributions to the Development of Iron Construction in the Nineteenth Century*. Essays in Public Works History, #21 (Kansas City, Mo., 2002).

## II. MILITARY

Browning, Robert S. *Two If By Sea: The Development of American Coastal Defense Policy* (Westport, Conn., 1983).

*Historical Papers Relating to the Corps of Engineers and to Engineer Troops in the Army. Occasional Paper No. 16*, United States Army Engineer School (Washington, D.C., 1904).

Lenney, John J. *Caste System in the American Army: A Study of the Corps of Engineers and the West Point System* (New York, N.Y., 1949).

Lewis, Emanuel Raymond. *Seacoast Fortifications of the United States: An Introductory History* (Washington, D.C., 1970).

Ludlow, William. "An Army Engineer's Journal of Custer's Black Hills Expedition, July 2, 1874–August 23, 1874." Edited by Eugene McAndrews. *Journal of the West* 13 (January 1974): 78–85.

Nelson, Harold L. "Military Roads for War and Peace, 1791–1836." *Military Affairs* 19 (Spring 1955): 1–14.

Robinson, Willard B. *American Forts: Architectural Form and Function* (Urbana, Ill., 1977).

## 1. American Revolution to the Mexican War

Beauregard, Pierre G.T. *With Beauregard in Mexico: Mexican War Reminiscences.* Edited by T. Harry Williams (Baton Rouge, La., 1956).

Guthorn, Peter J. *American Maps and Mapmakers of the Revolution* (Monmouth Beach, N.J., 1966).

Martin, Joseph Plumb. *A Narrative of Some of the Adventures, Dangers and Sufferings of a Revolutionary Soldier*. Edited by George F. Scheer (Boston, Mass., 1962).

Palmer, Dave Richard. *The River and the Rock: The History of Fortress West Point, 1775–1783* (New York, N.Y., 1969).

Robinson, William M., Jr. "The Engineer Soldiers in the Mexican War." *The Military Engineer* 24 (January–February 1932): 1–8.

Smith, Gustavus Woodson. *Company "A" Corps of Engineers, 1846–1848, in the Mexican War.* Edited by Leonne M. Hudson (Kent, Ohio, 2001).

Traas, Adrian G. *From the Golden Gate to Mexico City: The U.S. Army Topographical Engineers in the Mexican War* (Washington, D.C., 1993).

Walker, Charles E. "The Other Good Guys: Army Engineers in the War of 1812." *The Military Engineer* 70 (May–June 1978): 178–183.

Walker, Paul K. "An Engineering Victory: The Siege of Yorktown, 1781." *The Military Engineer* 73 (September–October 1981): 334–37.

Walker, Paul K. *Engineers of Independence: A Documentary History of the Army Engineers in the American Revolution, 1775–1783* (Washington, D.C., 1981).

## 2. Civil War to World War I

Cooling, Benjamin F. *Symbol, Sword, and Shield: Defending Washington During the Civil War* (Shippensburg, Pa., 1991).

*The Engineer Battalion in the Civil War. Occasional Paper No. 44*, United States Army Engineer School (Washington, D.C., 1910).

Floyd, Dale E., editor. *Dear Friends At Home: The Letters and Diary of Thomas James Owen, Fiftieth New York Volunteer Engineer Regiment, During the Civil War* (Fort Belvoir, Va., 1985).

Hendricks, Charles. *Combat and Construction: U.S. Army Engineers in World War I* (Fort Belvoir, Va., 1993).

Hotchkiss, Jedediah. *Make Me a Map of the Valley: The Civil War Journal of Stonewall Jackson's Topographer.* Edited by Archie P. McDonald (Dallas, Texas, 1973).

Knauff, Francis H. "Remember the Maine." *The Military Engineer* 47 (May–June 1955): 211–13.

Nichols. James L. *Confederate Engineers* (Tuscaloosa, Ala., 1957).

Thienel, Phillip M. "Engineers in the Union Army, 1861–1865." *The Military Engineer* 47 (January–February 1955): 36–41; (March–April 1955): 110–15.

Thienel, Phillip M. *Mr. Lincoln's Bridge Builders: The Right Hand of American Genius* (Shippensburg, Pa., 2000).

## 3. World War II

Anders, Leslie. *The Ledo Road* (Norman, Okla., 1965).

Beck, Alfred M., Abe Bortz, Charles W. Lynch, Lida Mayo, and Ralph F. Weld. *The Corps of Engineers: The War Against Germany.* United States Army in World War II: The Technical Services (Washington, D.C., 1985).

Coll, Blanche D., Jean E. Keith, and Herbert H. Rosenthal. *The Corps of Engineers: Troops and Equipment.* United States Army in World War II: The Technical Services (Washington, D.C., 1958).

Dod, Karl C. *The Corps of Engineers: The War Against Japan.* United States Army in World War II: The Technical Services (Washington, D.C., 1966).

Fine, Lenore, and Jesse A. Remington. *The Corps of Engineers: Construction in the United States.* United States Army in World War II: The Technical Services (Washington, D.C., 1972).

Fowle, Barry W., editor, *Builders and Fighters: U.S. Army Engineers in World War II* (Fort Belvoir, Va., 1992).

Fowle, Barry W., and Floyd D. Wright. *The 51st Again: An Engineer Combat Battalion in World War II* (Shippensburg, Pa., 1992).

Giles, Henry. *The G.I. Journal of Sergeant Giles* (Boston, Mass., 1965).

Giles, Janice Holt. *The Damned Engineers.* 2d ed. (Washington, D.C., 1985).

Heavey, William F. *Down Ramp! The Story of the Army Amphibian Engineers* (Washington, D.C., 1947).

Jones, Vincent C. *Manhattan: The Army and the Atomic Bomb.* United States Army in World War II: Special Studies (Washington, D.C., 1985).

Office of the Chief Engineer, General Headquarters, Army Forces Pacific. *Engineers of the Southwest Pacific, 1941–1945* (7 Volumes, Washington, D.C., 1947–1959).

Pergrin, Col. David E., with Eric Hammel. *First Across the Rhine: The 291st Engineer Combat Battalion in France, Belgium, and Germany* (New York, N.Y., 1989).

Twichell, Heath. *Northwest Epic: The Building of the Alaska Highway* (New York, N.Y., 1992).

Zarish, Joseph M. *The Collapse of the Remagen Bridge* (New York, N.Y., 1967).

## 4. Korean War and Vietnam Conflict

Dunn, Carroll H. *Base Development in South Vietnam, 1965–1970* (Washington, D.C., 1972).

Farquhar, William R., Jr., and Henry A. Jeffers, Jr. *Bridging the Imjin: Construction of the Libby and Teal Bridges During the Korean War.* Edited by Charles Hendricks (Fort Belvoir, Va., 1989).

Fowle, Barry W., and John C. Lonnquest, editors. *Remembering the "Forgotten War": U.S. Army Engineer Officers in Korea* (Alexandria, Va., 2005).

Ploger, Robert R. *U.S. Army Engineers, 1965–1970* (Washington, D.C., 1974).

Stearns, Maj. Henry A., Maj. Rudolph E. Abbott, and Capt. James L. Campbell. "LOC Highway Restoration, Vietnam." *The Military Engineer* 65 (March–April 1973): 84–86.

Strong, Paschal N. "Army Engineers in Korea." *The Military Engineer* 44 (November–December 1952): 405–10.

Westover, John G. *Combat Support in Korea: The United States Army in the Korean Conflict* (Washington, D.C., 1955).

Yens, 1st Lt. David P., and Capt. John P. Clement III. "Port Construction in Vietnam." *The Military Engineer* 59 (January–February 1967): 20–24.

### 5. Cold War and After

Grathwol, Robert P., and Donita M. Moorhus. *American Forces in Berlin, 1945–1994: Cold War Outpost* (Washington, D.C., 1994).

Grathwol, Robert P. and Donita M. Moorhus. *Building For Peace: U.S. Army Engineers in Europe, 1945–1991* (Washington, D.C., 2005).

Hatch, Lt. Gen. Henry J., and Janet A. McDonnell. "Corps of Engineers: Laying the Groundwork for Theater Operations." *Military Review* 72 (March 1992): 2–13.

McDonnell, Janet A. *After Desert Storm: The U.S. Army and the Reconstruction of Kuwait* (Washington, D.C., 1999).

McDonnell, Janet A. *Supporting the Troops: The U.S. Army Corps of Engineers in the Persian Gulf War* (Alexandria, Va., 1996).

Schubert, Frank N. *Building Air Bases in the Negev: The U.S. Army Corps of Engineers in Israel, 1979–1982* (Washington, D.C., 1992).

Schubert, Frank N. "U.S. Army Corps of Engineers and Afghanistan's Highways, 1960–1967." *Journal of Construction Engineering and Management* 117 (September 1991): 445–59.

Williams, Garland H. *Engineering Peace: The Military Role in Postconflict Reconstruction.* (Washington, D.C., 2005).

Winkler, David F., and John C. Lonnquest. *To Defend and Deter: The Legacy of the United States Cold War Missile Program* (Champaign, Ill., 1996).

## III. CIVIL WORKS

Anfinson, John O. "Commerce and Conservation on the Upper Mississippi River." *The Annals of Iowa* 52 (Fall 1993): 385–417.

Anfinson, John O. *The River We Have Wrought: A History of the Upper Mississippi* (Minneapolis, Minn., 2003).

Angevine, Robert G. *The Railroad and the State: War, Politics, and Technology in Nineteenth-Century America* (Stanford, Calif., 2004).

Arnold, Joseph L. *The Evolution of the Flood Control Act of 1936* (Fort Belvoir, Va., 1986).

Becker, William H. *From the Atlantic to the Great Lakes. A History of the U.S. Army Corps of Engineers and the St. Lawrence Seaway* (Fort Belvoir, Va., 1984).

Camillo, Charles A., and Matthew T. Pearcy. *Upon Their Shoulders: A History of the Mississippi River Commission From Its Inception*

*Through the Advent of the Modern Mississippi River and Tributaries Project* (Vicksburg, Miss., 2004).

Cowdrey, Albert E. "Pioneering Environmental Law: The Army Corps of Engineers and the Refuse Act." *Pacific Historical Review* 46 (August 1975): 331–49.

Fatherree, Ben H. *The First 75 Years: History of Hydraulics Engineering at the Waterways Experiment Station* (Vicksburg, Miss., 2004).

Ferejohn, John A. *Pork Barrel Politics: Rivers and Harbors Legislation, 1947–1968* (Stanford, Calif., 1974).

Ferrell, John R. *Big Dam Era: A Legislative and Institutional History of the Pick-Sloan Missouri River Basin Program* (Omaha, Neb., 1993).

Graves, Gregory. *Pursuing Excellence in Water Planning and Policy Analysis: A History of the Institute for Water Resources, U. S. Army Corps of Engineers* (Alexandria, Va., 1995).

Hagwood, Joseph J., Jr. *The California Debris Commission: A History of the Hydraulic Mining Industry in the Western Sierra Nevada of California, and the Governmental Agency Charged with Its Regulation* (Sacramento, Calif., 1981).

Hill, Forrest G. *Roads, Rails, and Waterways: The Army Engineers and Early Transportation* (Norman, Okla., 1957).

Holland, Francis Ross, Jr. *America's Lighthouses: Their Illustrated History Since 1716* (Brattleboro, Mass., 1972).

Holmes, Beatrice Hort. *A History of Federal Water Resources Programs, 1800–1960* (Washington, D.C., 1972).

Holmes, Beatrice Hort. *History of Federal Water Resources Programs and Policies, 1961–70* (Washington, D.C., 1979).

Johnson, Leland R. "Army Engineers on the Cumberland and Tennessee, 1842–1854." *Tennessee Historical Quarterly* 31 (Summer 1972): 149–69.

Johnson, Leland R. "Waterways: The Fourth Pillar of Defense." *The Military Engineer* 72 (November–December 1980): 404–08.

Maass, Arthur. *Muddy Waters: The Army Engineers and the Nation's Rivers* (Cambridge, Mass., 1951).

Mazmanian, Daniel A., and Jeanne Nienaber. *Can Organizations Change: Environmental Protection, Citizen Participation and the Corps of Engineers* (Washington, D.C., 1979).

Merritt, Raymond H. *The Corps, The Environment, and The Upper Mississippi River Basin* (Washington, D.C., 1981).

Moore, Jamie W., and Dorothy P. Moore. *The Army Corps of Engineers and the Evolution of Federal Flood Plain Management Policy* (Boulder, Colo., 1989).

Morgan, Arthur E. *Dams and Other Disasters: A Century of the Army Corps of Engineers in Civil Works* (Boston, Mass., 1971).

Nichols, Roger L. "Army Contributions to River Transportation, 1818–1825." *Military Affairs* 33 (April 1959): 242–49.

Pearcy, Matthew T. "After the Flood: A History of the 1928 Flood Control Act." *Journal of the Illinois State Historical Society* 95 (Summer 2002): 172–201.

Pearcy, Matthew T. "A History of the Ransdell-Humphreys Flood Control Act of 1917." *Louisiana History* 41 (Spring 2000): 133–59.

Petersen, Keith C., and Mary E. Reed. *Controversy, Conflict, and Compromise: A History of the Lower Snake River Development* (Walla Walla, Wash., 1994).

Power, Garrett. "The Fox in the Chicken Coop: The Regulatory Program of the U.S. Army Corps of Engineers." *Virginia Law Review* 63 (1977): 503–59.

Reuss, Martin. "Andrew A. Humphreys and the Development of Hydraulic Engineering: Politics and Technology in the Army Corps of Engineers, 1850–1950." *Technology and Culture* 26 (January 1985): 1–33.

Reuss, Martin. "The Army Corps of Engineers and Flood-Control Politics on the Lower Mississippi." *Louisiana History* 23 (Spring 1982): 131–48.

Reuss, Martin. "Coping with Uncertainty: Social Scientists, Engineers, and Federal Water Planning." *Natural Resources Journal* 32 (Winter 1992): 101–35.

Reuss, Martin. *Designing the Bayous: The Control of Water in the Atchafalaya Basin, 1800–1995* (Alexandria, Va., 1998).

Reuss Martin. "Reshaping National Water Politics: The Emergence of the Water Resources Development Act of 1986." *IWR Policy Study* 91-PS-1 (Fort Belvoir, Va., 1991).

Reuss, Martin. *Shaping Environmental Awareness: The United States Army Corps of Engineers Environmental Advisory Board, 1970–1980* (Washington, D.C., 1983).

Shallat, Todd A. "Building Waterways, 1802–1861: Science and the United States Army in Early Public Works." *Technology and Culture* 31 (January 1990): 18–50.

Shallat, Todd A. *Structures in the Stream: Water, Science, and the Rise of the U.S. Army Corps of Engineers* (Austin, Texas, 1994).

Shallat, Todd A. "Water and Bureaucracy: Origins of the Federal Responsibility for Water Resources, 1787–1838." *National Resources Journal* 32 (Winter 1992): 5–25.

Stine, Jeffrey K. *Mixing the Waters: Environment, Politics, and the Building of the Tennessee-Tombigbee Waterway* (Akron, Ohio, 1993).

Stine, Jeffrey K. "The Tennessee-Tombigbee Waterway and the Evolution of Cultural Resources Management." *The Public Historian* 14 (Spring 1992): 6–30.

U.S. Army Corps of Engineers. *National Waterways Roundtable Papers: Proceedings on the History and Evolution of U.S. Waterways and Ports* (Fort Belvoir, Va., 1981).

Walker, Paul K. "Building American Canals." *Water Spectrum* 12 (Winter 1979–80): 18–25; (Summer 1980): 12–23.

Willingham, William F. "Engineering the Cascades Canal and Locks, 1876–1896." *Oregon Historical Quarterly* 88 (Fall 1987): 229–57.

Wood, Lance D., and John R. Hill, Jr. "Wetlands Protection: The Regulatory Role of the U.S. Army Corps of Engineers." *Coastal Zone Management Journal* 4 (1978): 371–407.

## IV. DISASTER RESPONSE

Burgess, Carter L. "The Armed Forces in Disaster Relief." *The Annals of the American Academy of Political and Social Science* 309 (January 1957): 71–79.

Cooling, B. Franklin. "The Army and Flood and Disaster Relief." *The United States Army in Peacetime: Essays in Honor of the Bicentennial, 1775–1975*. Edited by Robin Higham and Carol Brandt (Manhattan, Kan., 1978): 198–200.

Deakyne, Herbert. "Bridging Kaw River in the 1903 Flood." *The Military Engineer* 20 (May–June 1928): 198–200.

Johnson, Leland R. "19th Century Engineering: The Johnstown Disaster." *The Military Engineer* 66 (January–February 1974): 42–45.

Johnson, Leland R., "No Work, No Rations: Army Engineers' Work Relief in Southern States." *Journal of Mississippi History* 67 (Summer 2005): 107–33.

McCullough, David G. *The Johnstown Flood* (New York, N.Y., 1968).

McDonnell, Janet A. *Response to the Loma Prieta Earthquake* (Fort Belvoir, Va., 1993).

McDonnell, Janet A. *The U.S. Army Corps of Engineers Response to the* Exxon Valdez *Oil Spill* (Fort Belvoir, Va., 1992).

Sieg, Kent G. "Lessons Learned in Response and Recovery: The U.S. Army Corps of Engineers at the World Trade Center." *Joint Center for Operational Analysis and Lessons Learned Bulletin* (December 2004): 36–43.

Walker, Paul K. *The Corps Responds: A History of the Susquehanna Engineer District and Tropical Storm Agnes* (Baltimore, Md., 1976).

Willingham, William F. "The Army Corps of Engineers' Short-Term Response to the Eruption of Mount St. Helens." *Oregon Historical Quarterly* 106 (Summer 2005): 174–203.

## V. SURVEYS AND EXPLORATIONS

Baldwin, Kenneth H. *Enchanted Enclosure: The Army Engineers and Yellowstone National Park: A Documentary History* (Washington, D.C., 1976).

Bartlett, Richard A. *Great Surveys of the American West* (Norman, Okla., 1962).

Beers, Henry P. "History of the U.S. Topographical Engineers, 1813–1863." *The Military Engineer* 34 (June 1942): 287–91; (July 1942): 348–52.

Fielding, Robert K. *The Unsolicited Chronicler: An Account of the Gunnison Massacre, Its Causes and Consequences* (Brookline, Mass., 1993).

Goetzmann, William H. *Army Exploration in the American West 1803–1863* (New Haven, Conn., 1959).

Nichols, Roger L., and Patrick L. Haley. *Stephen Long and American Frontier Exploration* (Norman, Okla., 1995).

Ronda, James P. *Beyond Lewis and Clark: The Army Explores the West* (Tacoma, Wash., 2003).

Simpson, James H. *Navaho Expedition: Journal of a Reconnaissance from Santa Fe, New Mexico, to the Navaho Country, Made in 1849.* Edited by Frank McNitt (Norman, Okla., 2003).

Schubert, Frank N. *The Nation Builders: A Sesquicentennial History of the Corps of Topographical Engineers, 1838–1863* (Fort Belvoir, Va., 1988).

Schubert, Frank N. *Vanguard of Expansion: Army Engineers in the Trans-Mississippi West, 1819–1879* (Washington, D.C., 1980).

Wallace, Edward S. *The Great Reconnaissance: Soldiers, Artists and Scientists on the Frontier, 1848–1861* (Boston, Mass., 1955).

## VI. AUTOBIOGRAPHY AND BIOGRAPHY

Bishop, Joseph B., and Farnham Bishop. *Goethals: Genius of the Panama Canal, A Biography* (New York, N.Y., 1930).

Brodhead, Michael J. "James W. Abert: Soldier, Explorer, and Naturalist." *Journal of America's Military Past* 27 (Spring-Summer 2000): 65–75.

Chaffin, Tom. *Pathfinder: John Charles Frémont and the Course of Empire* (New York, N.Y., 2002).

Clark, Edward B. *William L. Sibert, The Army Engineer* (Philadelphia, Pa., 1930).

Dodds, Gordon B. *Hiram Martin Chittenden: His Public Career* (Lexington, Ky., 1973).

Franzwa, Gregory M., and William J. Ely. *Leif Sverdrup, Engineer Soldier at His Best* (Gerald, Mo., 1980).

Freeman, Douglas Southall. *Lee* (New York, N.Y., 1997).

Gifford, Emerson. *Gouverneur Kemble Warren: The Life and Letters of an American Soldier, 1830–1882* (Boston, Mass., 1932).

Haiman, Miecislaus. *Kosciuszko in the American Revolution* (New York, N.Y., 1943).

Heusser, Albert H. *George Washington's Map Maker: A Biography of Robert Erskine.* Edited by Hubert G. Schmidt (New Brunswick, N.J., 1966).

Humphreys, Henry H. *Andrew Atkinson Humphreys, A Biography* (Philadelphia, Pa., 1924).

Jordan, David M. *"Happiness Is Not My Companion": The Life of General G. K. Warren* (Bloomington, Ind., 2001).

Kite, Elizabeth S. *Brigadier-General Louis Lebègue Duportail* (Baltimore, Md., 1933).

Lewis, Gene D. *Charles Ellet, Jr.: The Engineer as Individualist, 1810–1862* (Urbana, Ill., 1968).

McAndrews, Eugene V. "Custer's Engineer—William Ludlow." *The Military Engineer* 61 (May–June 1969): 200–02.

Miller, David W. *Second Only to Grant: Quartermaster General Montgomery C. Meigs, A Biography* (Shippensburg, Pa., 2000).

Mumey, Nolie. *John Williams Gunnison (1812–1853): The Last of the Western Explorers* (Denver, Colo., 1955).

Nevins, Allan. *Fremont: Pathmarker of the West* (2 Volumes, New York, N.Y., 1961).

Nichols, Maj. Gen. K.D., U.S.A. (Ret.). *The Road to Trinity: A Personal Account of How America's Nuclear Policies Were Made* (New York, N.Y., 1987).

Norris, L. David, James C. Milligan, and Odie B. Faulk. *William H. Emory: Soldier-Scientist* (Tucson, Ariz., 1998).

Norris, Robert S. *Racing for the Bomb: General Leslie R. Groves, the Manhattan Project's Indispensable Man* (South Royalton, Vt., 2002).

Richards, Kent D. *Isaac I. Stevens: Young Man in a Hurry* (Provo, Utah, 1979).

Sauers, Richard A. *Meade: Victor of Gettysburg* (Washington, D.C., 2003).

Sears, Stephen W. *George B. McClellan: The Young Napoleon* (New York, N.Y., 1988).

Smith, Jean Edward. *Lucius D. Clay* (New York, N.Y., 1990).

Snell, Mark A. *From First to Last: The Life of Major General William B. Franklin* (New York, N.Y., 2002).

Weigley, Russell F. *Quartermaster General of the Union Army: A Biography of M.C. Meigs* (New York, N.Y., 1959).

Whaley, Elizabeth M. *Forgotten Hero: General James B. McPherson: The Biography of a Civil War General* (New York, N.Y., 1955).

Williams, Thomas Harry. *P.G.T. Beauregard: Napoleon in Gray* (Baton Rouge, La., 1955).

Wood, Richard G. *Stephen Harriman Long, 1784–1864: Army Engineer, Explorer, Inventor* (Glendale, Calif., 1966).

## VII. ORAL HISTORY INTERVIEWS

*Colonel William W. Badger*. Interviewed by Frank N. Schubert (Washington, D.C., 1983).

*Major General Hugh J. Casey*. Interviewed by John T. Greenwood (Washington, D.C., 1993).

*Lieutenant General Frederick J. Clarke*. Interviewed by Albert E. Cowdrey, Suellen Hoy, and Michael C. Robinson (Washington, D.C., 1979).

*Lieutenant General Carroll H. Dunn*. Interviewed by Paul K. Walker (Washington, D.C., 1998).

*William R. Gianelli*. Interviewed by Martin Reuss (Washington, D.C., 1985).

*Lieutenant General Ernest Graves*. Interviewed by Frank N. Schubert (Washington, D.C., 1997).

*General William M. Hoge*. Interviewed by Maj. Gen. George Robertson (Washington, D.C., 1993).

*Major General Richard S. Kem.* Interviewed by William C. Baldwin, Paul K. Walker, and John T. Greenwood (Alexandria, Va., 2002).

*Arthur Maass.* Interviewed by Martin Reuss (Washington, D.C., 1989).

*Frederick McNeely.* Interviewed by Frank N. Schubert (Washington, D.C., 1987).

*Lieutenant General John W. Morris.* Interviewed by William C. Baldwin (Alexandria, Va., 2000).

*Major General William E. Potter.* Interviewed by Martin Reuss (Washington, D.C., 1983).

*Lieutenant General Edward O. Rowny.* Interviewed by Barry W. Fowle (Washington, D.C., 1995).

*Lieutenant General Arthur G. Trudeau.* Interviewed by Col. Calvin J. Landau (Washington, D.C., 1986).

*Gilbert F. White.* Interviewed by Martin Reuss (Washington, D.C., 1993).

*Lieutenant General Walter K. Wilson, Jr.* Interviewed by Paul K. Walker (Washington, D.C., 1984).

# Index

## A



## B

## F



## G

## H

## M

## S

## Y